T0145293

A Water Grid for England

An Alternative View of Water Resources in England

PETER STYLES

AuthorHouse™ UK
1663 Liberty Drive
Bloomington, IN 47403 USA
www.authorhouse.co.uk
UK TFN: 0800 0148641 (Toll Free inside the UK)
UK Local: 02036 956322 (+44 20 3695 6322 from outside the UK)

All of the diagrams are by the author.

Thanks again to Wikipedia, Wikimedia, Creative Commons and all of the free sites who make knowledge sharing so easy these days.

This book is printed on acid-free paper.

ISBN: 978-1-7283-7944-9 (sc)
ISBN: 978-1-7283-7943-2 (e)

Print information available on the last page.

Published by AuthorHouse 12/05/2022

authorHOUSE®

Dedicated to all who take an alternative view to that held by the mainstream.

Also by Peter Styles
(sometimes writing as Stilovsky and Schrodinger and sometimes as himself)

Hoggrills End **published December 2017**

The Power of Numbers **published January 2018**

The Power of Names **published May 2018**

The Power of Notes **published September 2018**

The Power of Words **(1) published December 2018**

Power Quiz '18 **published January 2019**

Power Quiz '19 **published March 2019**

Power Quiz '17 **published July 2019**

The Power of Words (2) **published March 2020**

The Power of Words (3) **published May 2020**

The Power of Dreams **published July 2020**

SAMS, Simplified Asset Management Systems **published December 2020**

Principles of Asset Management **published 2021 with Wayne Earp**

The Power of Water **published 2022**

Please be aware that the author has no connection with the
American gay romance author of the same name.

[all available in paperback or Kindle on Amazon]

Contents

We forget that the water cycle and the life cycle are one

Jacques Eves Cousteau

Foreword

In the absence of a formal foreword I have reproduced some words by Dame Margaret Beckett when she introduced the recent report* on infrastructure resilience (or rather - lack of it).

> "There are plenty of examples of the extremely serious impact that climate change has already had on our critical national infrastructure. And there are bound to be more in the future – almost certainly more serious still.
>
> But the thing I find most disturbing is the lack of evidence that anyone in Government is focusing on how all the impacts can come together, creating cascading crises. There are simply no ministers with focused responsibility for making sure that our infrastructure is resilient to extreme weather and other effects of climate change.
>
> Storm Arwen showed how quickly the effects of a power shutdown can impact on other sectors. People were left without any access to their landline phones after the storms, and unable even to call emergency services in areas with a poor mobile signal. These cascading crises are a major danger to the functioning of the UK economy, and to society – that's why this is a serious risk to national security.
>
> Events such as Storm Arwen and the summer heat wave are going to happen more and more often. We heard just last week that the UK may face blackouts early next year if we lose further gas supplies from Europe. The new Prime Minister must pull all the strands of government together to mitigate against potential disasters, including climate change impacts. This Government must finally recognise that prevention is better than cure and move on from their dangerously reactive approach to risk management."

* *Readiness for storms ahead? Critical national infrastructure in an age of climate change.*
[Published by The Joint Committee on the National Security Strategy, October 2022]

Introduction and Background

Utilities are those private and government bodies that provide us with the essentials for life and include:

- Roads and Railways
- Postal services
- Power – electricity and gas
- Communications – telephone, radio and television
- Water – supply and sewerage

A short perusal of these industries is enough to show that, with the exception of water, they all have national grids so the service is not interrupted except in exceptional circumstances. Remember that many of our early motorways were constructed as by-passes and only joined up later. The route of the M6 by-passed Stafford, Preston and Lancaster before it was linked. And, even then, the route through Birmingham took years to complete - what would things be like without 'Spaghetti Junction' today?

Canals were once the 'A Roads' of their day but they were all built by the private sector who took a limited view of connecting them. Gas Street Basin in Birmingham was a major bottleneck and it wasn't until someone had the bright idea of building the Grand Union (the M1 of its day) that the canals became a national network. Is history is repeating itself?

The first postal services in the UK, of the 18th century, were based in townships which then joined up to involve five separate postal companies. They were amalgamated in 1840 to become the Uniform Penny Post and eventually Royal Mail.

A similar problem occurred in the early days of building the railways but it soon became apparent that a national network was the answer. Whilst the canals were concentrated on connecting industrial areas, London was always the focus for railways, as it has become for HS2.

The gas system was built by local government and fragmented when we used town gas because it was generated locally. Once we had North Sea gas a national network was rolled out in very short time. Electricity was generated and supplied locally until we had the National Grid.

Getting back to the subject - the risks to our water supply include:

- Climate change and drought
- Pollution
- Infrastructure failure
- Stress due to over-use
- Increasing population

Whilst weather is a major player, the possibilities of deliberate interference must not be overlooked.

Firstly, Scotland, Northern Ireland and Wales are self sufficient as is much of Northern England and the West as this is where the rainfall is sufficient to cover demand even in a drought.

Secondly, let's dispel any notion that a water grid would operate like the national grid – moving potable water around the country under pressure. If a grid is proposed, then its primary purpose would be to address the raw water deficits especially in the South and South East of England. It would consist of inter-regional transfers of raw water which would be pumped or gravitated, between reservoirs and rivers.

Having been a water engineer for over forty years, in Local Government, a major water company and then as an international consultant, I have always had an interest in water matters and am now trying to get some clarity into an issue which has dogged the industry for decades. Since Carsington Reservoir was commissioned in 1991, there has not been a major development of water resources in England. The current mode of regulation, and the presence of the water-only companies, makes any form of joined-up thinking difficult.

But a long career in the industry does not keep you up-to-date so I must apologise if any of my stuff is past its sell-by-date. Also, as much of the detail of water systems is omitted, there are many minor facilities which are not included since they play little part in the national picture. I have not included any details of aquifers.

Some will complain that I have not numerically described the problems or costed them. As far as the first is concerned – there are many reports out there which describe the problem in numerical terms. In my view, they tend to obscure the nature of the problem and provide an excuse for 'playing at trains'. Costing is essential but cost/benefit is difficult when you are dealing with a matter of national importance. The Joint Committee on National Security do not delve into the minutiae of the problems which are being created by climate change and neither will I. Apologies over.

A water grid has been talked about for many years and some saw privatisation as the key that would unlock the formula to bring one about. Unfortunately, the mode of regulation that has been imposed on the industry has made this impossible as the economic regulator gives no brownie points for helping your neighbour. The environmental regulator appears to have taken up the mantle of promoting water transfers in England in the absence of anyone with a clear mandate from government to pursue this agenda.

The current organisation which has been set up to promote water resources across England still suffers from parochial interests. Five regional groups have been set up, overseen by the Environment Agency (EA) which has attempted to bring things together in a series of reports produced in the Spring of 2022. Whilst terms of reference were supplied to the groups (Appendix 1), they appear to have paid little heed to them.

- WReN (Water Resources North)
- WRW (Water Resources West)
- WCWR (West Country Water Resources)
- WRE (Water Resources East)
- WRSE (Water Resources South East)

Over the last forty years, little has been done to ensure that no part of England will run dry in a severe drought. The 2022 drought brought things into focus but now that the rain is falling again, there is little sign of progress.

The five regional groups have just published their 'final' reports (November 2022). The EA criticised the groups for their lack of consistency and inability to address this issue in their draft reports but I still see little in the way of inter-regional transfers. The problem is endemic in the industry model which has ten large water and sewerage companies plus a number of smaller water-only companies which were exempt from the 1974 reorganisation. Little has changed since other than privatisation of the larger companies. No employee of one of these smaller companies is going to agree with any proposal which might make them dependent on one of their larger neighbours and hence makes them liable to takeover. If you doubt that this is a possibility – just Google 'East Worcestershire Water Company'.

To illustrate the problem, let's consider one of the five interim reports – WReN (Water Resources North) which covers Kielder in the north to Sheffield in the south. The report is factual, detailed and considers a number of proposals including several possible regional transfers. Each is taken in isolation as a stand-alone project without any consideration of the overall picture or even the national context. This bottom-up approach is OK when you are concerned solely with local matters but it's short-term and lacks joined-up thinking. Using a top-down approach, each of the projects can be seen in a national context and viewed from a long-term perspective – how does it fit in with the national strategy and will if still be useful in 2072?

Whilst the National Infrastructure Commission (NIC) includes discussion on the issue, it has little, if anything, in the way of concrete proposals. So, my paper (if that's what it is) is a top-down look at water resources, in contrast to the bottom-up approaches of the five regional groups et al.

You may disagree with what I have published but ask yourself – why? In 1986, the development of a strategic grid was proposed to link all of the major water treatment plants in the West Midlands so that they could support eachother in times of stress or crisis. This was greeted with derision by the assistant directors with related departments who saw the plan as an infringement on their areas of responsibility. It took some time to work out that this was NIHS* at work. Following the Worcester/Wem incident, the proverbial hit the fan and this was followed by flooding of The Mythe. There was much confusion during both incidents and the company was fined. When a single department took over from the previously fragmented teams, common sense prevailed. It took 25 years to complete the grid but it's all there exactly as proposed. Most other major companies have followed suit.

*Not invented here syndrome (look it up on Wikipedia)

It's very tempting to say "I told you so" but that's not the point. We have to get out of our silo mentality and look at the big picture instead of spending all of our time manipulating numbers and immersing ourselves in detail.

Since The Water Resources Board was disbanded, no government body or organisation has taken on the mantle of planning for water resources on a national scale. The Environment Agency has played at it with the setting up of the five regional groups which were tasked with bringing a national plan together. Some would describe this as fragile as they have produced inconsistent and over-detailed reports to date. Their 'final' reports were published in November of 2022 but there is little sign of a national strategy to come from the exercise.

The Environment Agency appears ambivalent about its role in the exercise and has settled on trying to cajole the groups to see the bigger picture. The groups have, in compiling their reports, basically concentrated on internal issues with only scant regard for inter-regional transfers. The EA's comments on the draft reports of January 2022, criticises this major omission in a very restrained manner.

Only a few worthwhile projects have emerged alongside a plethora of minor internal projects and some vague suggestions. The Severn-Thames Transfer (STT) at Deerhurst is eminently sensible though the project appraisal is still in its early stages. The new reservoir near Abingdon (called SESRO for some strange reason) is also essential though Thames appear reluctant to get on with it. The Mendip Quarries scheme is an excellent example of what can be done but its inter-linking is only on a minor scale. Anglian Water have just announced the sites for two new reservoirs which are isolated from their other assets.

Ofwat's 'Gated System of Project Appraisal' resembles a job creation scheme and is so complicated that it will only hold up schemes rather than promote them. Little progress seems to have been made with RAPID (a misnomer if ever there was one) though the groups' final reports have moved things on – if only a little.

The primary options for creating a reliable, future-proofed resource system for England are:

- It's raining so why worry?
- Rely on reducing demand and leakage
- Provide limited connectivity with a view to regular review
- Create a Water Grid as part of a long-term National Strategy
- Create more recycling schemes
- Rely on desalination
- Any of the above in combination

Water reuse is already widely practised through abstraction from rivers which have treated effluent in them at low concentrations. A colleague told me, some years ago, that Rutland Water was largely filled with Peterborough's effluent. Whether we are able to extend this and make more use of the abundant flow in some under-used rivers remains to be seen though there has been little mention of it to date.

Let's look at some basics. Here's a theoretical catchment:

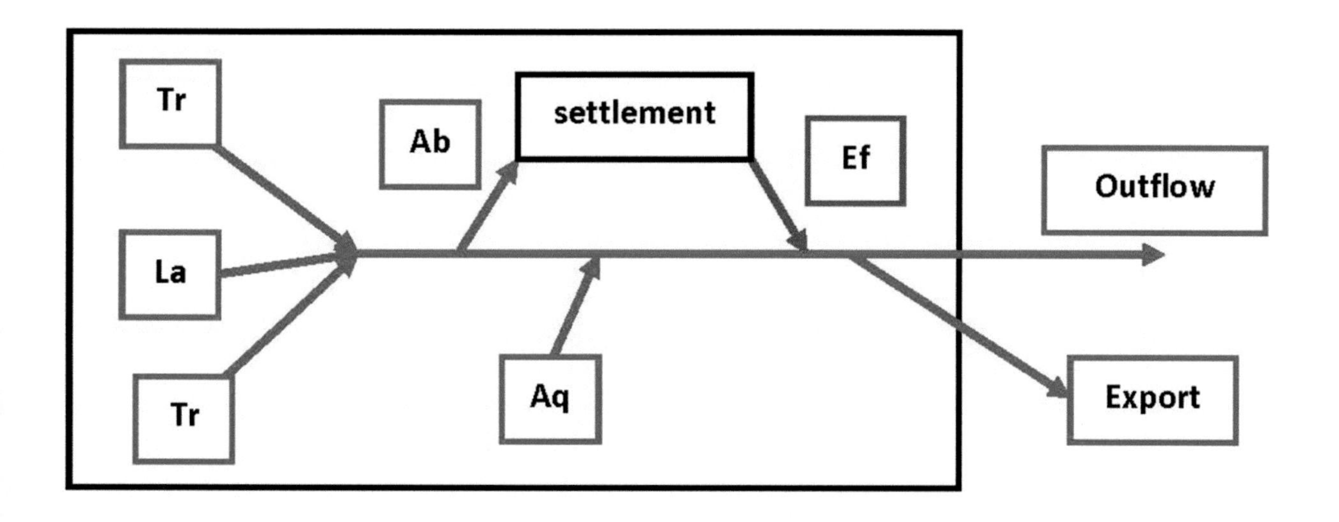

In simple terms, the inflows to the river are from the tributaries plus whatever comes directly from the aquifer (Aq). All of these have one source – rainfall which falls on the catchment - less evaporation. Where a catchment is self contained then any settlements have a neutral effect on the total resource, as abstraction (Ab) is almost equal to the effluent return (Ef). However, paved areas will speed up the response to rainfall and reduce the inflow to the aquifer. The current concentration on SuDS is intended to mitigate both of these. If an export is taken from the catchment (such as at Vyrnwy) then that water is lost. Lakes do not fundamentally change the overall resource – they slow everything down so the flow in the river is balanced and hence proceeds more slowly. Reservoirs are man-made lakes which are managed to store water and release it in a controlled manner – either to the river or to direct abstraction.

Analysing the bodies of surface water (streams, rivers, lakes and reservoirs) is relatively simple but analysing an aquifer is difficult as it lies out-of-sight beneath the surface. There is also a difference, between the shallow aquifers which lie close to water courses and the deep sandstone and limestone aquifers. The former tend to lie in gravel beds and the water is similar, if a little cleaner, than that in the river. In contrast the water contained in deep aquifers tends to be cleaner and suitable for direct supply if just chlorinated. The shallow beds are easily restored by the river or groundwater but the deep aquifers need substantial rainfall.

Whilst the effect of abstracting water direct from a river, lake or reservoir is immediately apparent, sinking a borehole into a deep aquifer needs to be carefully and continually monitored. A water balance is needed to ensure longevity of the aquifer. Once the water table sinks below the ground level, then a spring at that point will dry up.

Why am I writing this basic stuff? Well, if you're a water engineer it's all old hat but if you're a politician, you may be thinking – why don't we just tap another aquifer? That's what they did in California and just look at where it's got them. At least we have an understanding of our resources and control of them.

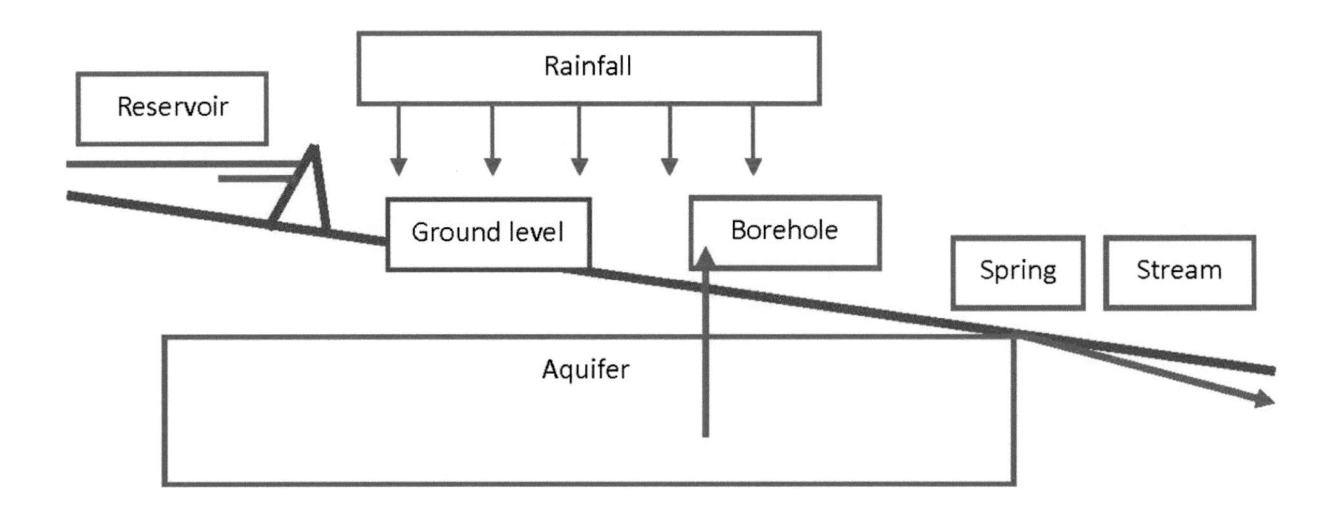

If we want to look at it in personal finance terms, it would be a little like this. A savings account (like an aquifer) is a finite resource and depleted when a withdrawal is made. Our current account (a river) is topped up with our income (rainfall) and depleted by spending (abstractions). The balance in our current account flows on. Excess income (rainfall) might find its way into our savings account but the path is more complex unless we have a standing order.

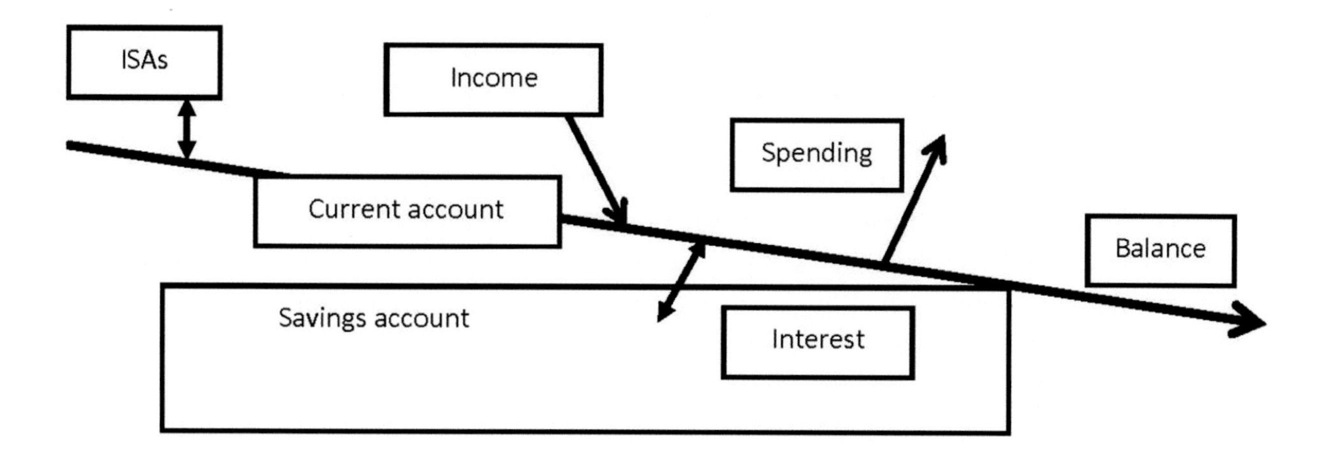

I'm sure you can make a better version of this and even include your credit card?

Stating the Obvious

1. Scotland, Northern Ireland and Wales do not have the same resource problem that is apparent in England
2. Any resource problems in Wales are confined to the South
3. The North and West of England have few resource issues and they can be managed within their own boundaries using their internal water grids
4. The South and South East of England, including London are in serious deficit amounting to some 4,000 Ml/d in 2050; The published percentages for deficits are: 47% South East; 25% East; 17% West; 6% West Country and 5% North
5. There is no lack of treatment capacity in any region – the problems relate to water resources
6. Reducing leakage will not help - most leakage simply recharges the local aquifer meaning that it is not lost as a resource and any reduction in individual consumption will be offset by population increase and may not, therefore, be relied on to reduce consumption overall*
7. It is cheaper, using membrane technology, to treat sewage effluent to a potable standard than saline or brackish waters
8. When a drought occurs, it is not uniform across the country which makes inter-regional transfers more resilient than local solutions
9. We are happy to consume treated wastewater so long as it has sufficient dilution and we don't question exactly where it's come from
10. Any advanced treatment, such as Reverse Osmosis (RO), is expensive and has a high carbon footprint meaning that it should be avoided on cost and carbon deficit grounds
11. Aquifers are finite; drawing from them is not a sustainable option as it permanently depletes resources
12. Artificial aquifer recharge is costly as it requires treatment of the raw water in excess of potable standards
13. Any proposal for a raw water grid should utilise existing assets wherever possible
14. The closing of the Water Resources Board, along with the National Water Council, has left a void in the system for resource planning which the Environment Agency struggles to fill
15. Current reports by the five regional groups are concentrated on detail and lack conformity or any sense of a clear direction in a national context
16. Under the current regulatory regime, water companies see no benefit in engaging on inter-regional raw water transfers which, in turn, means that they try to resolve any resource issues within their own boundaries
17. Water companies have little incentive to consider other demands on water resources, in particular agriculture and wetlands
18. Canals are not designed to transfer water and designs for such will be problematical.

* WRSE plan says "Leakage reduction and demand management measures are core to our overall strategy. They will deliver 70% of the overall solution in the first five years of the plan and remain at over 50% of the solution by the end of the planning period."

(IMHO, this is highly unlikely to be achieved and largely irrelevant anyway)

The Current Situation and Regional Proposals

General

Most of the large water companies already have an internal strategic grid to move treated water around their region though there is little in the way of transfers between neighbours. These potable water links are hardly relevant to the planning of a water grid which is concerned with raw water.

Recent discussions in New Civil Engineer suggested a spine, gravity-operated aqueduct running down the Pennines supplying lower lying areas along the route. It would, presumably, terminate on Cannock Chase where a number of Midlands systems could be supplied by gravity. This would operate as a linear asset, taking water in just one (north to south) direction which could be transferring either raw or potable water. It would not solve the problem in the South.

That parts of the grid, that inter-link existing reservoirs, could look like this:

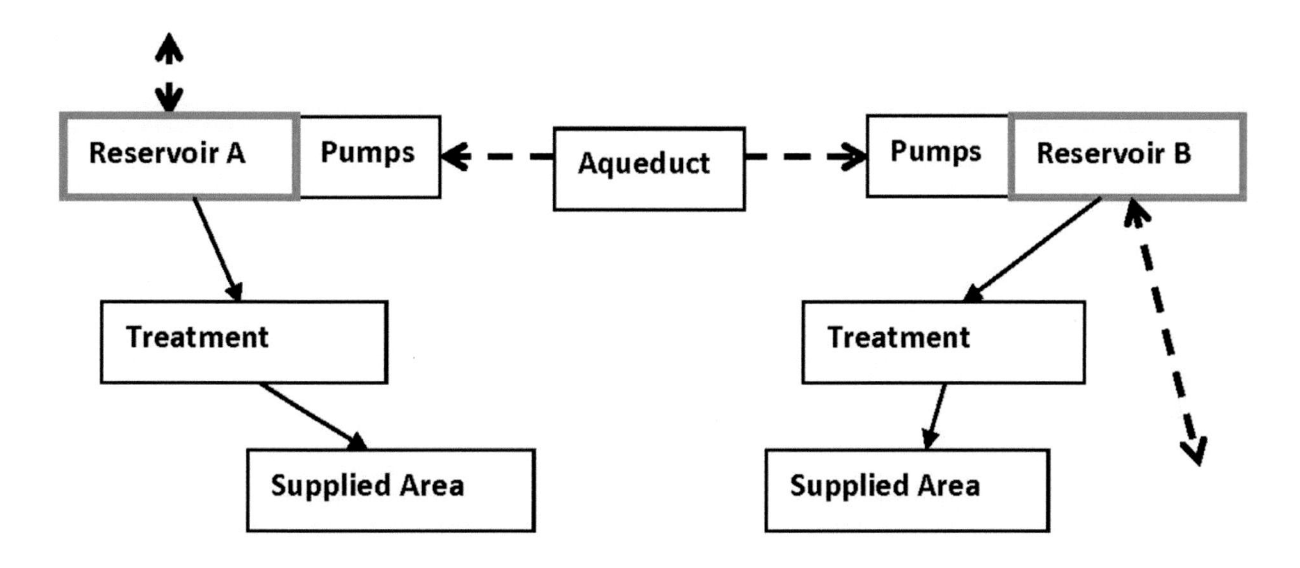

A raw water grid would use existing major reservoirs and link them with a series of aqueducts which are able to operate in either direction – thus providing mutual support. The pump stations and pipelines would be built to a common standard and a limited range of pipe diameters would be used (say 750 or 1,000mm diameter).

Kielder Water is a major resource which provides guaranteed supplies to the whole of Northumbria as far south as the River Tees which is connected by a raw water aqueduct. Both the reservoir and the pipeline were subject to much debate and appraisal before construction and the system is somewhat underused due to the decline in heavy industry in Newcastle, Sunderland and Middlesbrough. A question arises whether there should be any interlinking between the Kielder system and the aqueducts which serve Manchester?

Ladybower Reservoir serves both the East Midlands (Severn Trent) and Sheffield (Yorkshire Water). The joint supply agreement will run out in 2030 and it is possible that the supply will be directed south to fill any deficit resulting from restrictions on abstraction from the Nottinghamshire sandstones. This will, in turn, create a new demand in South Yorkshire which will need to be fed from the North. It would follow that it would make sense to bring Kielder water further south by extending the aqueduct which currently terminates at the River Tees. This could then release water from the northern Yorkshire reservoirs to be brought further south.

If this water is available in South Yorkshire, it is a relatively short step to connect it up with Carsington which would form a base for distribution further south. The question, then, is whether to connect Carsington directly with Rutland Water or proceed via the Leicestershire reservoirs? A direct link could have a branch to Foremark.

With the East Midlands, which already has a sophisticated system to move water around, we need to consider whether the Trent could be incorporated. This river has a very consistent dry weather flow, as it contains the treated effluent from Minworth (Birmingham), Strongford (Stoke), Wanlip (Leicester), Sponden (Derby) and Stoke Bardolph (Nottingham). As industry has declined, and treatment standards have risen, water quality has improved and the closure of many of the power stations has brought the temperature back closer to normal. Obviously a major investigation is necessary but one, carried out some years ago by Severn Trent, indicated that groundwater in the Trent Valley was suitable for treatment and hence supply. A 'sough' running in parallel with the river could provide an inexhaustible resource of relatively high quality raw water.

Links from the Midlands to the East and the South would link up the major reservoirs though some transfer could be possible using canals.

There are currently four new major reservoirs planned:

- The planning for Mendip Quarries is at an advanced stage
- The planning for SESRO near Abingdon, to regulate the River Thames, is also advanced
- The site for the Lincolnshire reservoir is now decided
- The site for the Fens reservoir is now decided

Plans to support the Thames catchment include transfers from the River Severn (STT) and using the Grand Union Canal plus a new reservoir near Abingdon (SESRO). If the Thames catchment is seen to be in surplus, then raw water could be transferred south. A minor reservoir is planned by Portsmouth Water at Havant Thicket and it is planned to support it using treated sewage effluent. Whilst the South is in serious deficit, little else is planned in the way of major projects in the near future.

The new reservoir at Mendip Quarries has minor links. Kent stands on its own and would benefit from inter-linking of the five main reservoirs. Any additional resource would probably come from blending treated wastewater in Bewl Water.

What follows is a more detailed consideration of each region.

The North East

Kielder Water is the major resource in the whole of the North East of England. It already feeds an aqueduct which takes raw water south to support the River Tees. The Eastern Pennines boast a number of reservoirs which benefit from mutual support. The system based on Scammonden suffered a severe shortage some years ago and water was brought in by tankers at great expense. The Ladybower sharing agreement with Severn Trent will expire in 2030 so something needs to be done to service the needs of Sheffield. If Scammonden is linked with the Howden reservoir in the Derwent Valley, the agreement should be able to continue. However this could trigger a need to support Scammonden from the Vales Reservoirs which might, in turn mean bringing water south from the River Tees system. Thus, a raw water grid would enable a transfer from the north further south. (*note – the two 'Derwents' have nothing to do with eachother)

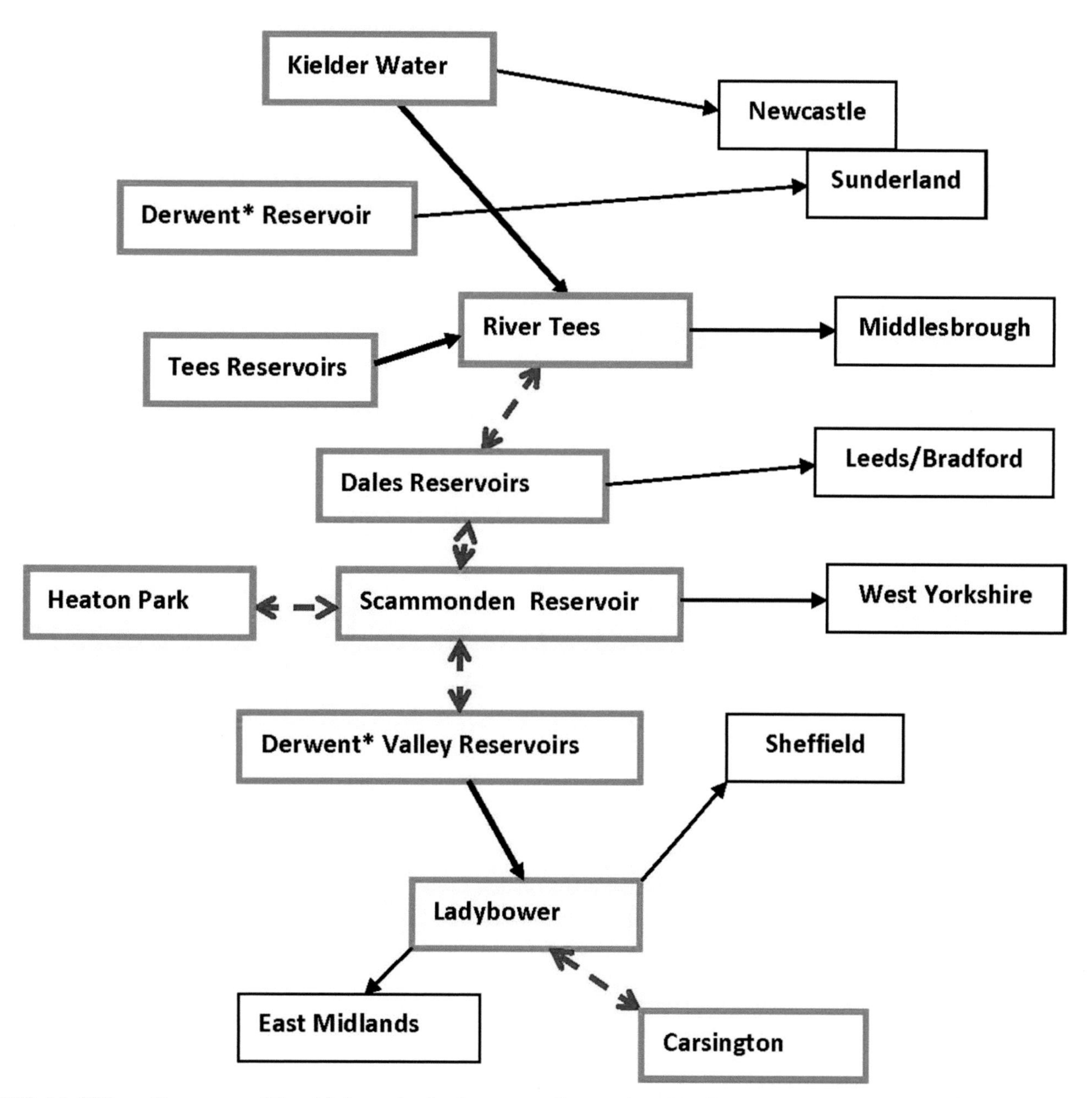

WReN (Water Resources North) have looked at transfers to WRW (Water Resources West) from Kielder but I have considered Scammonden to Heaton Park as a more practical option.

The North West

There are two principle water supply systems serving the North West; having been developed by Manchester and Liverpool Corporations respectively. There is also the Lancashire Conjunctive Use Scheme which is designed to transfer raw water from the River Lune at Caton to the River Wyre where it can be extracted for treatment and used to supply South Lancashire. It is not significant in national terms. These systems have an interconnection, for maintenance purposes, but there is no connection with the North East system.

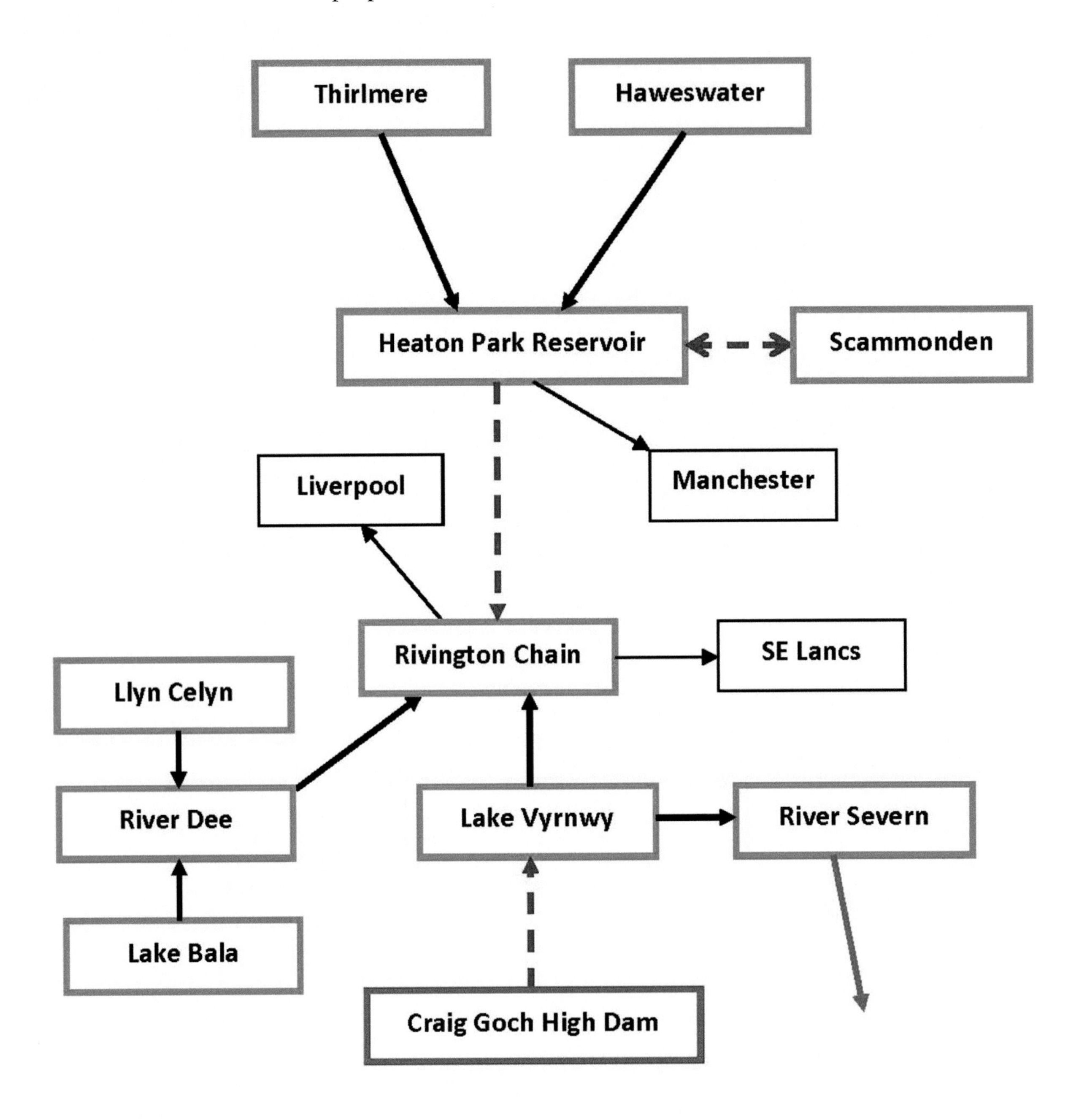

The Midlands

There are two principle water supply systems serving the East and West Midlands plus South Staffs and the Potteries. The only link between the East and West Midlands systems provides potable water from the East to Oldbury Reservoir which serves Nuneaton. The Potteries have a conjunctive use system which utilizes reservoirs under normal conditions and ground water when it's dry. South Staffs have a conjunctive use system and potable water connections with Severn Trent. Wolverhampton relies largely on groundwater.

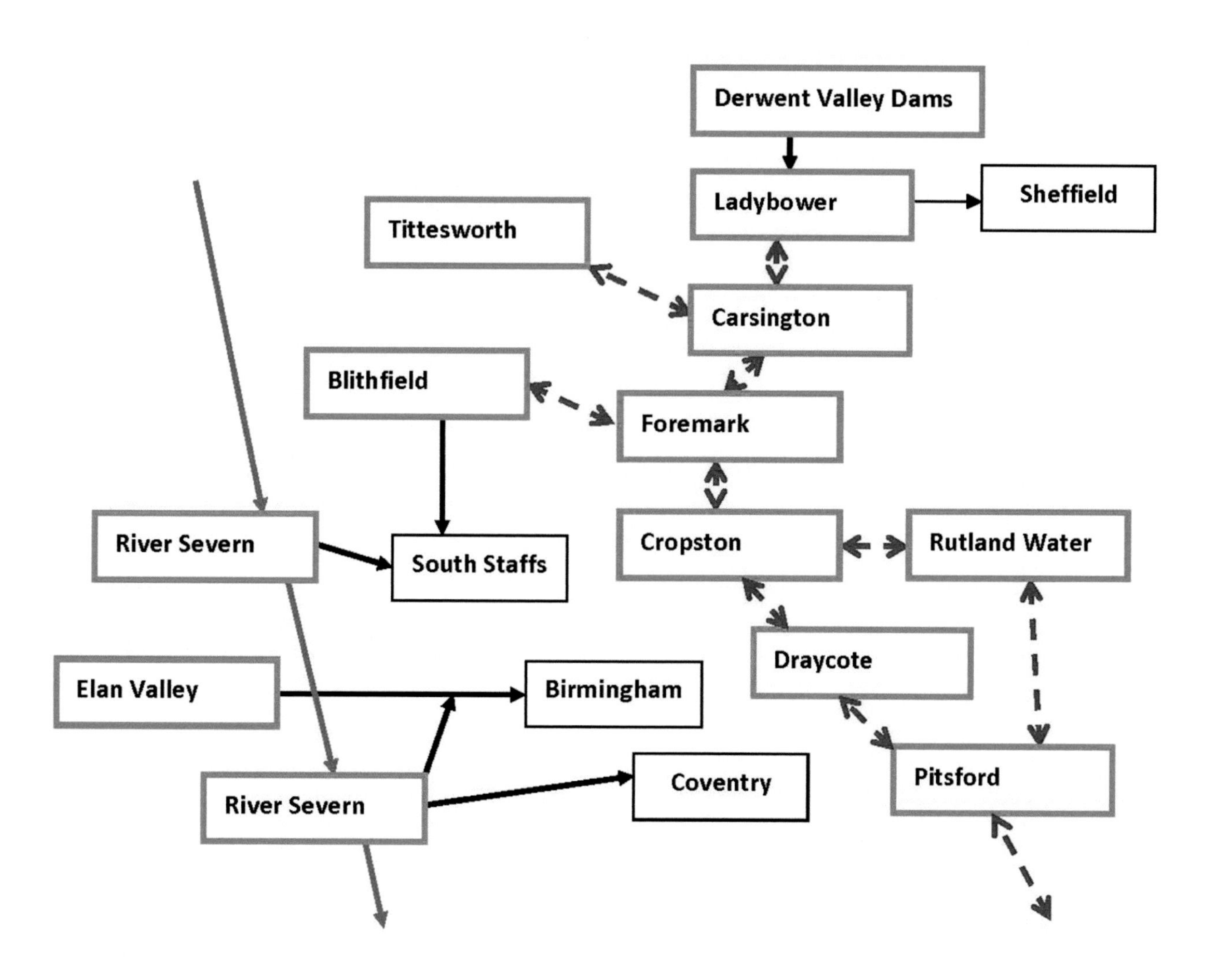

Severn Trent has two water treatment plants near the confluence of the Warwickshire Avon with the Severn – Strensham, which serves Coventry and The Mythe which serves the local area. I mention this as they are within spitting distance of the proposed Severn Thames Transfer (STT) location at Deerhurst. The initial project appraisal (or lack of it) for STT failed to mention this. Would it be possible to use either, or both, of these plants to transfer raw or treated water into the Thames catchment via Bredon Hill?

The East

The Eastern Reservoirs, proposed by Anglian Water have recently become firm proposals as the sites have now been chosen. Both should be considered alongside water transfers from the existing major reservoirs in the area.

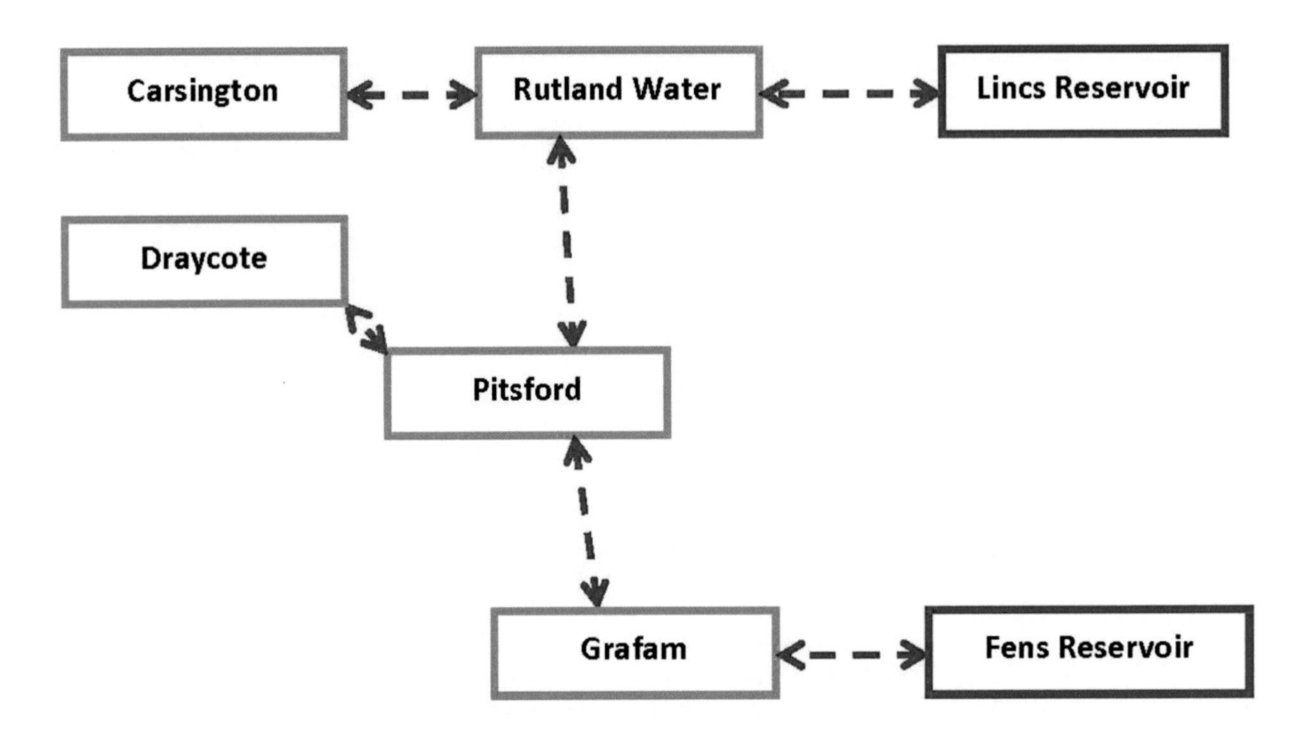

Hanningfield and Abberton reservoirs could be interlinked and fed with treated Beckton effluent.

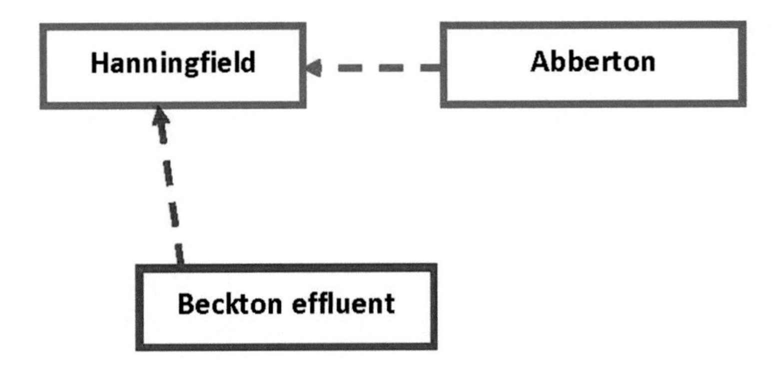

The West Country

The Mendip Quarries Reservoir proposal is good but limited in effect as it has not been connected with other reservoirs in the West Country. The picture is complicated as there are four main water supply companies involved: South West Water; Wessex Water; Bristol Water and Bournemouth Water. Despite its proximity, Severn water is not available in this area as the Bristol Channel is saline.

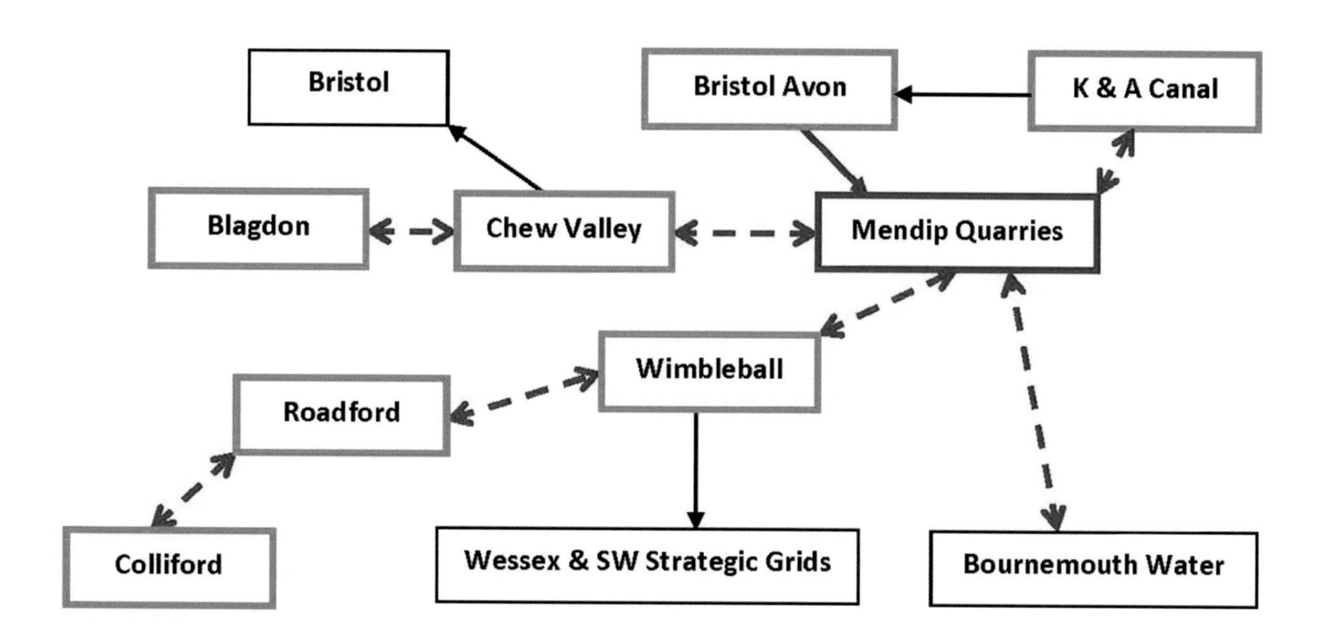

The Mendip Quarries Scheme is undergoing planning and appraisal at present and construction will commence when the quarries cease to be in use. This project provides the opportunity to look at the wider context and envisage what a support grid could look like if the resources were under the control of a single agency. The proposed link to the Kennet and Avon Canal should connect with the summit pound of the canal. This would make a self-contained system as the water from the summit pound gravitates down the canal to the Bristol Avon where it can be abstracted to Mendip Quarries. This enables the canal to cease its draw on the Thames catchment.

Direct pumping from the Bristol Avon to the summit pound would appear to be a more obvious solution?

The Thames Catchment

The River Thames is the most heavily abstracted river in England and has a serious resource deficit. It is also strategically important as it provides most of the water supplied to the capital.

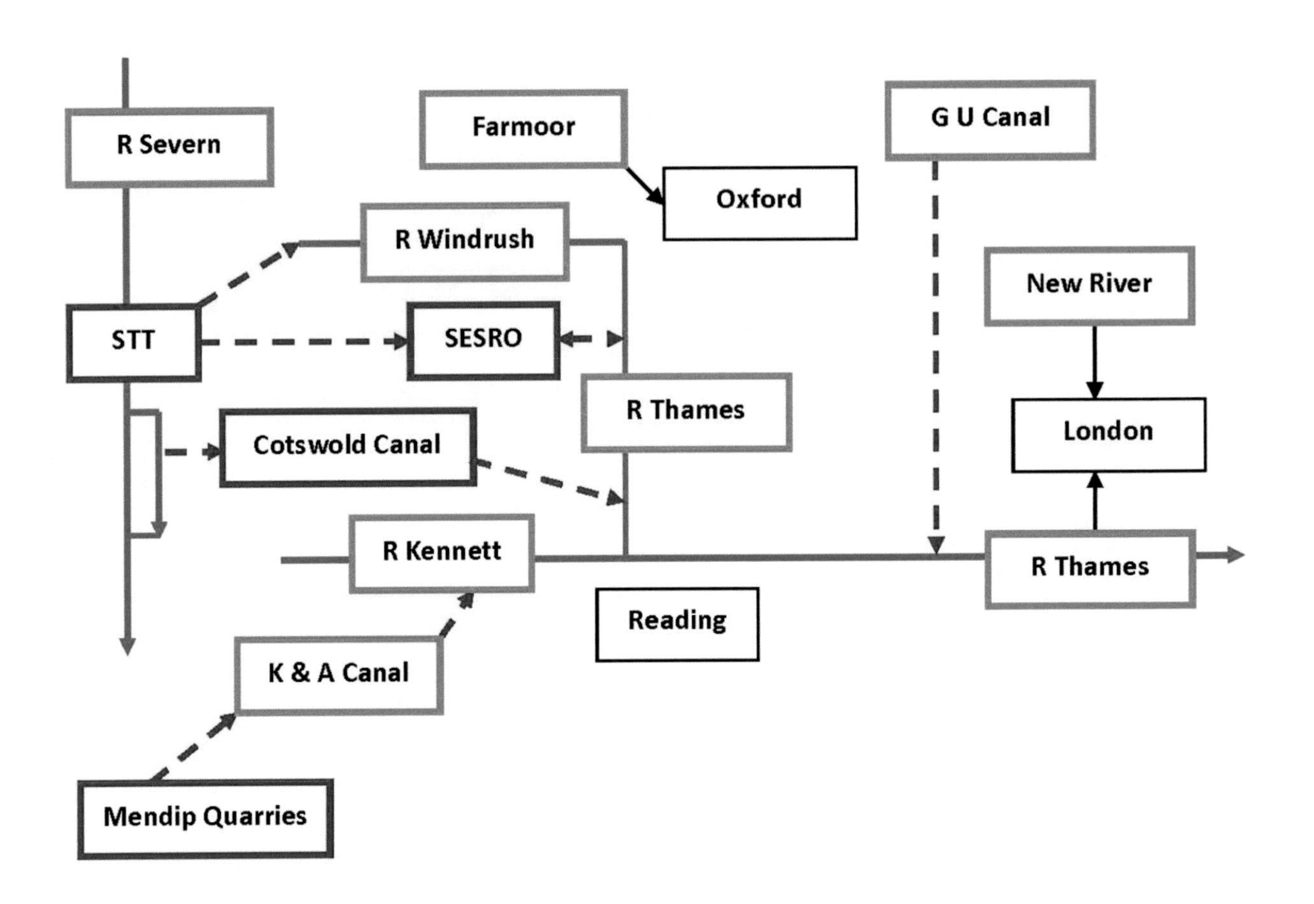

[note SESRO is "South East Strategic Resource Option" – wouldn't "Abingdon Reservoir" be easier?]

The Kennet & Avon Canal currently creates a drain on the Thames catchment but this could be corrected if the proposed Mendip Quarries reservoir were to supply water to the summit pound near Marlborough.

The Severn – Thames Transfer (STT) scheme could provide water from the Severn into the headwaters of the Thames catchment but still has to undergo detailed project appraisal and design. Whilst shown in the diagram above, the scheme is actually very much more complex. The current proposal appears to compare a direct pipeline with the Cotswold Canal as a means to transfer 300Ml/d directly into tributaries of the River Thames. Direct transfer needs careful consideration as it would materially change the nature of the Cotswold streams, one of which flows through Bourton-on-the-Water. In addition, it is highly unlikely that the Cotswold Canal can be seriously considered as a comparable option but it could be considered as a pipeline route. The canal itself may not be linked up for at least 20 years.

Issues at the Deerhurst end of the STT are a fairly simple matter of design: intake; screening; filters and pipes. However, once the water is transferred into the Thames catchment, there are a number of fundamental issues to consider. The proposal was to let the pre-treated raw water flow out into the Thames tributaries which will cause problems itself if 300Ml/d is released in a single or just a few locations.

It is proposed to build a new reservoir alongside the Thames near Abingdon (SESRO) which would be fed by pumping and hence regulate the river. At least some of the STT water could be taken directly there by pipeline.

One of the problems in the upper reaches of the catchment is aquifer depletion and this is made clear as the Thames itself no longer rises at its historic source near Cricklade. Restoring the aquifers in the catchment would provide a healthier river system than feeding raw water directly into streams. Three options spring to mind: injection via boreholes, spray irrigation and drip irrigation. Supplying raw water to farms would also reduce the demand on the local streams and help restore the aquifers without the need for additional management. The discharge of raw waters into the Windrush and the Leach requires careful consideration and the best solution overall is likely to include all of the above in order to distribute the catchment loading as widely as possible.

The South Eastern group's failure to consider whether The Mythe and Strensham could be used to supply treated water into the Thames catchment, via Bredon Hill, is indicative of the silo mentality which pervades this subject. I have commented on it elsewhere.

Overall the position for the Thames catchment can be summarised:

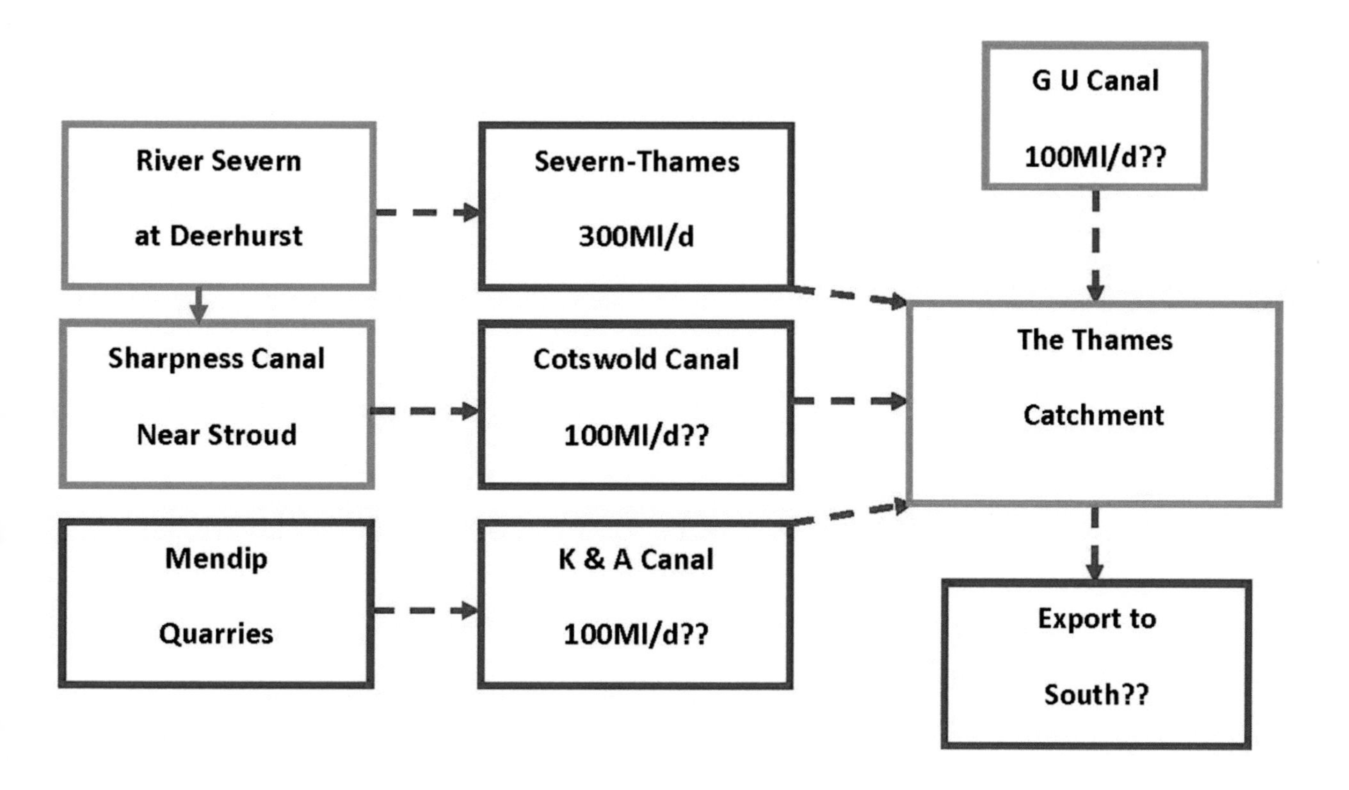

The South

Whilst the Thames catchment is included in the remit of the WRSE group, I have considered it separately. The South Coast of England is problematic as there are no major reservoirs in the whole area and the Thames is the only major river which borders it. The catchments, which contain chalk streams, are in serious deficit and need to be reinforced as further development of groundwater sources will only make things worse. The problem is exacerbated by the existence of the number of water-only companies which creates a silo mentality.

The Havant Thicket Reservoir/reuse scheme is at an advanced stage but is of limited scale in the regional context. It does, however, provide a sound model for wastewater recycling.

Any link from the River Thames/SESRO would have to come after the Thames catchment is augmented.

The Ardingly and Mendip Quarries links are possible but unlikely.

The links between the Bournemouth, Southampton and Portsmouth systems are only likely to come about following a major amalgamation of the companies concerned. As all three border onto the English Channel, other options are available. Whilst desalination is under consideration, it could be as cost effective to bring in short-term imports of fresh water during periods of drought. Kielder Water, shipped via Newcastle, has been considered in the past - this could be raw or potable water.

The long term, future-proofed solution would be to construct a new major reservoir in the South Downs National Park which could be linked with Ardingly Reservoir in the Kent system.

The three companies would obviously benefit from the development of a strategic grid, linking their potable systems.

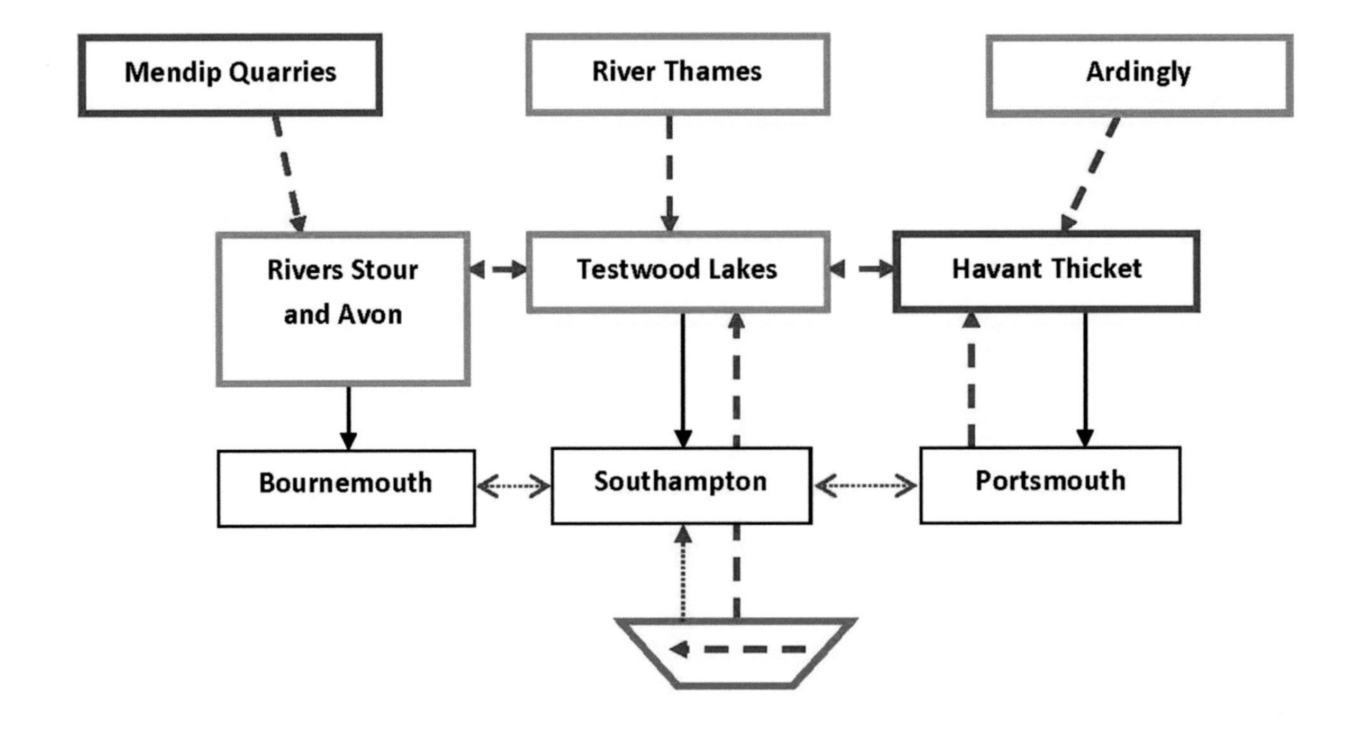

The report of the WRSE group contains a spider diagram which shows proposals for the interlinking of systems but it is not apparent whether it refers to raw water or potable supplies. In the latter case this would be a 'strategic grid' and have little to do with the resource strategy.

Kent

Despite being surrounded by water on three sides, Kent is virtually an island in water resource terms and is, hence, difficult to link into any national plan. It does, however, have on its doorstep, an inexhaustible supply of raw water which is sufficient to ensure the supply of the whole county. This requires that we address 'the yuck factor' as it involves recycled effluent from Crossness or elsewhere.

The focus of any development of a grid for Kent must start with its biggest resource - Bewl Water.

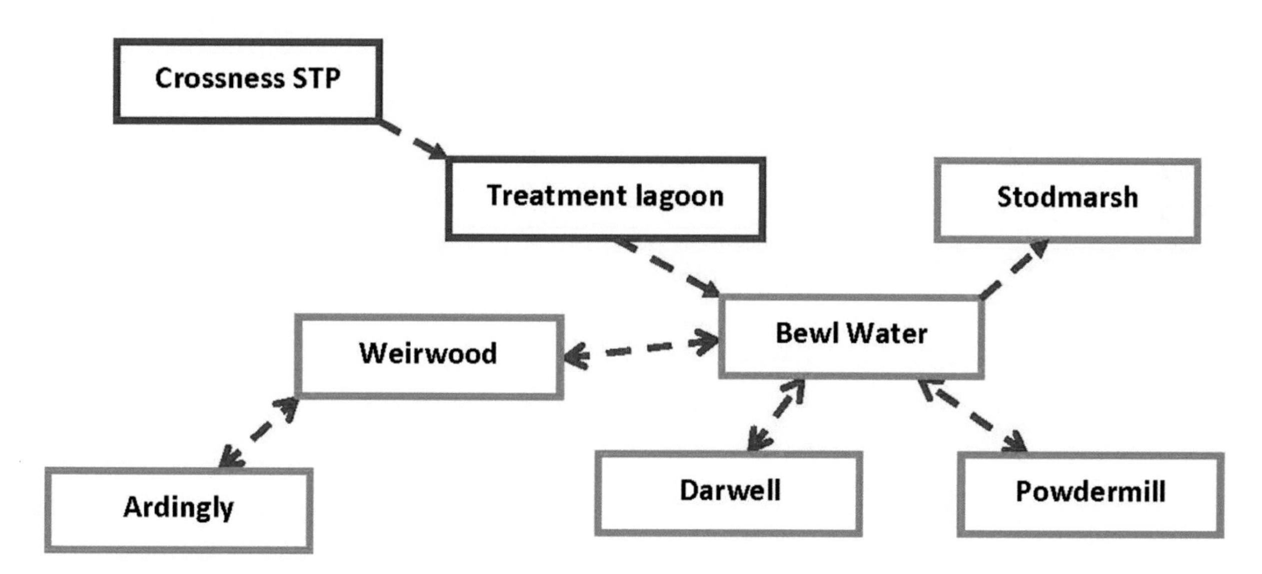

Crossness effluent currently flows into the Thames Estuary but it could be treated to a higher standard in two phases. First it would need to be brought up to discharge standards as if it were discharged to an inland watercourse, which could be achieved on the existing site. Secondly, the effluent would need to be 'polished' and this might require a new treatment lagoon to be constructed along the route between the plant and the reservoir.

Crossness is some considerable distance from Bewl Water so Maidstone and/or Tunbridge effluent could be a cheaper option. The required treatment system would be similar to that proposed for Havant Thicket in Hampshire i.e. sand filters.

The Kent Raw Water Grid would link the reservoirs together to provide mutual support in times of stress.

Canals

There are at least three proposals for raw water transfers using canals.

- The Grand Union provides a route between the Midlands and the South. This could transfer water southwards and two proposals have already been made to use Minworth effluent
- The Kennet and Avon links the Severn Estuary near Bristol with the Thames catchment near Devizes
- The Cotswold Canal has only been partly restored and a substantial gap still exists. This is being considered as a route for the STT scheme along with the route from Deerhurst on the Lower Severn

They have limited value as a constant flow will tend to hinder navigation and increase erosion of the banks. The Canal and River Trust are likely to require substantial guarantees and this is will be costly.

The Grand Union Canal

There has been much talk about using the Grand Union Canal as a conduit to feed raw water south – in the direction of London. A short perusal of the facts shows that this is not going to be easy.

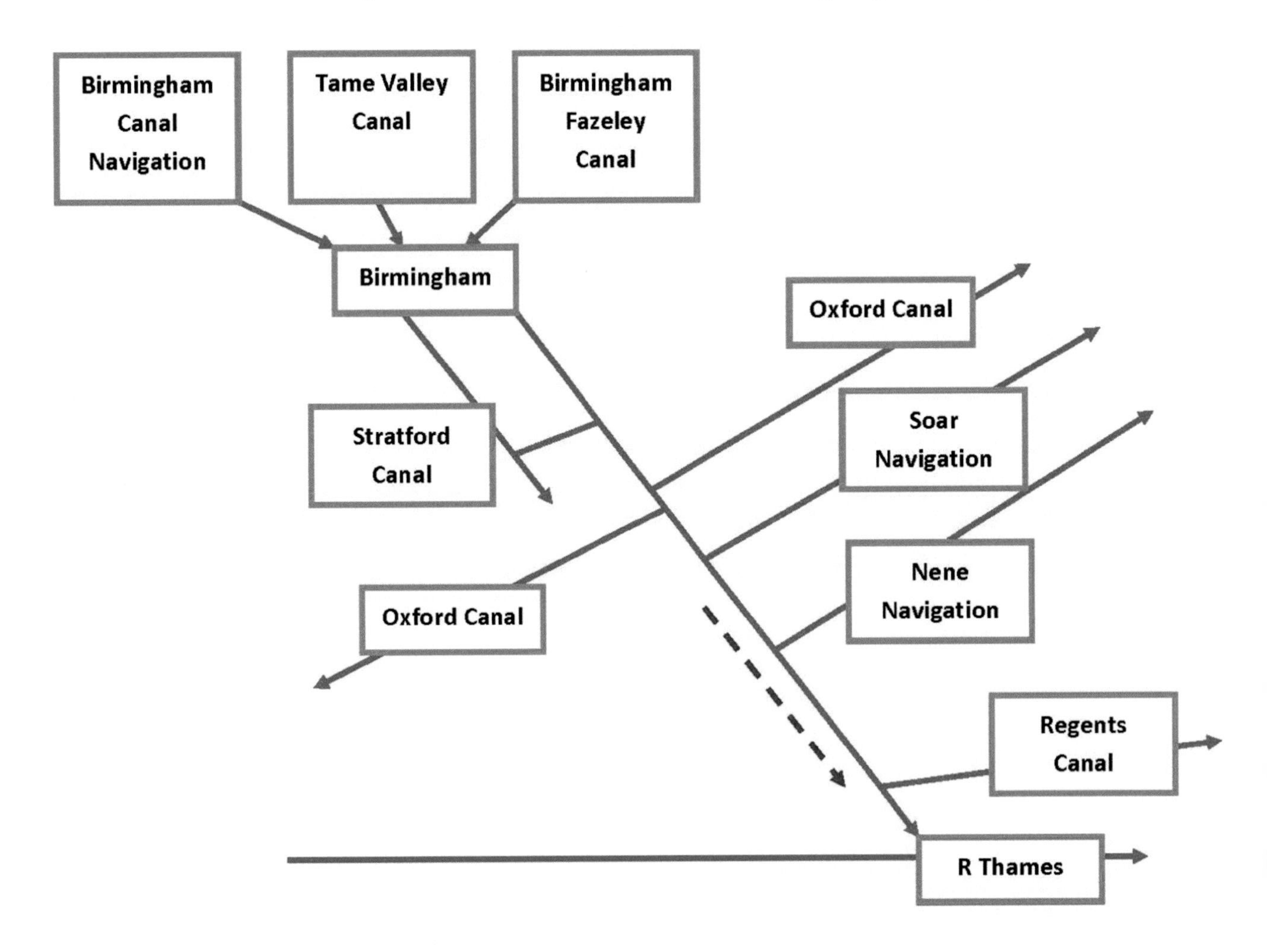

Whilst Birmingham is situated on the Midlands Plateau at around 300ft and London at near sea level, the canal does not have a steady incline in its fall of 280ft (86m). It has three summit pounds along its length.

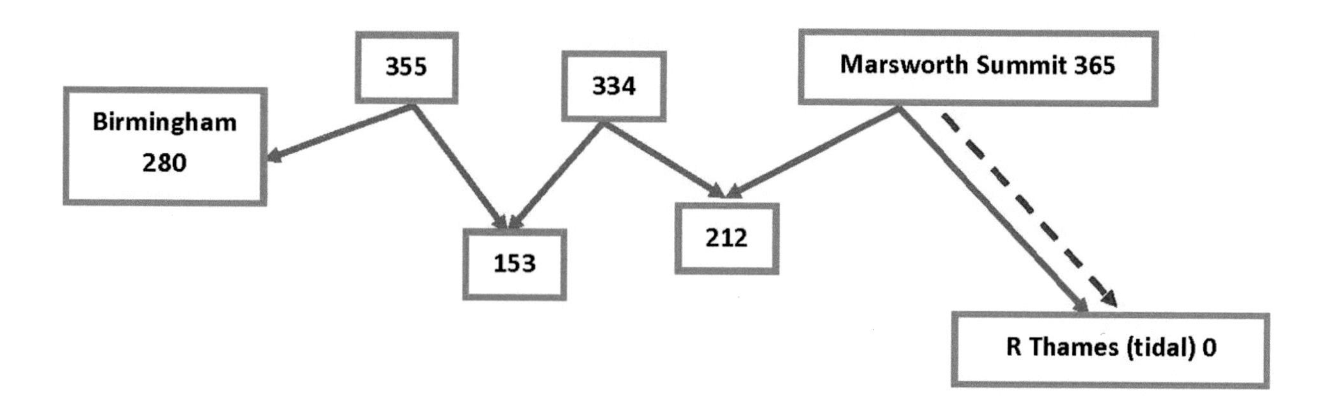

This means that any scheme which goes beyond Marsworth Summit pound would be complex and require multi stage pumping. Even the proposal to use the canal to carry Minworth effluent to the River Avon would require pumping. I feel that it is unlikely that any canal will be required to carry more than 100Ml/d as greater quantities would hinder navigation and could cause scouring. Even this could be problematic in by-passing the locks which abound on the canal network.

The Kennet & Avon Canal

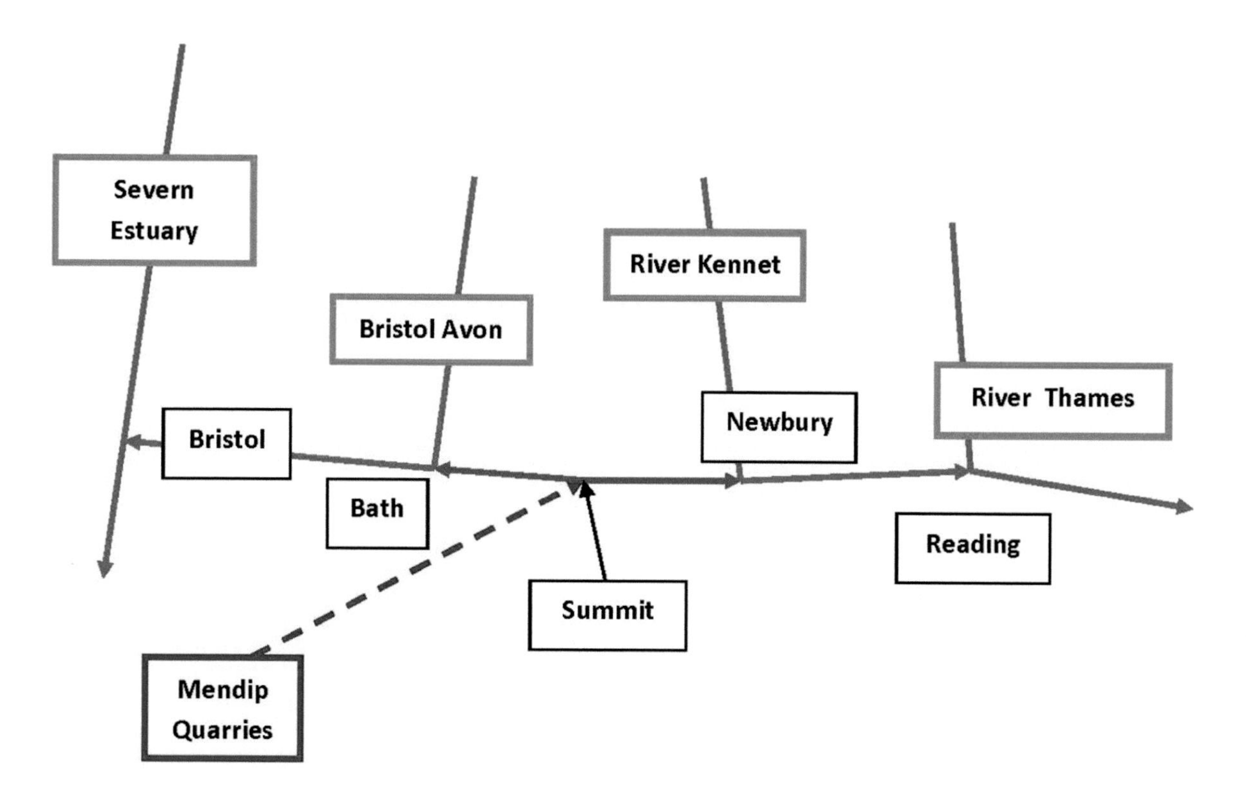

Whilst the connection between the two rivers (the Bristol Avon and the Kennet) is short, its climb over the summit involves a large number of locks. There are three pumping stations on the canal which replenish water lost as boats pass through the locks. Any provision of water from the proposed Mendip Quarries Reservoir would need to reach at least to the summit pound in order to gravitate into the Thames catchment. It would, however, be simpler to pump water from the Sharpness Canal direct to the summit pound where it would gravitate into the Thames catchment. This could be considered as an option to the Deerhurst proposal for STT.

The Cotswold Canal

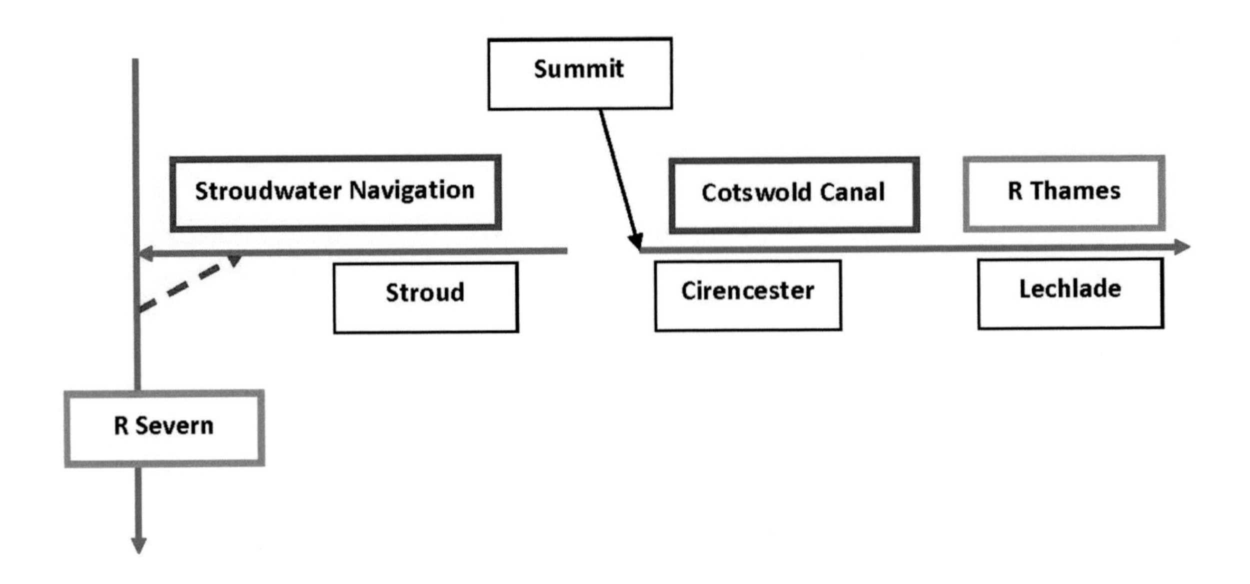

The canal joins the River Severn as the Stroudwater Navigation which was built first to connect Stroud with the Severn Estuary. The rest of the canal was built to join Stroud to the Thames at Lechlade and hence provide a route between the Severn and London via the Thames. It was largely superseded when the Kennet & Avon Canal was opened. It is now owned by The Cotswold Canal Trust who are attempting to join up the gaps. Their planning horizon is about 20 years hence and is heavily dependent upon National Lottery funding. When completed, the summit pound, if serviced from the Thames catchment, would make further demands on a depleted resource.

As a carrier for water from the Severn to the Thames, it does not provide a viable solution. The route could, however, be considered for a pipeline between Stroud and Cirencester which is just beyond the summit pound meaning that water would flow into the Thames catchment. Abstraction could be from the Sharpness Canal which already supplies Bristol.

The Rivers of Central England and Wales

The Severn

When I joined the water company, someone explained that we treat the Severn with respect as she's the source of much of our good water. But we treat the Trent as a sewer – to take away all the bad stuff from Birmingham, the Black Country and the Potteries.

Appendix 2 shows the River Severn in resource terms. It is the most heavily used resource of all the rivers except the Thames, with the exception that its management has ensured that it is a reliable resource into the distant future. Even the proposed abstraction of 300Ml/d for the STT can be managed though the detail has yet to be decided.

My own choice would be to build the Craig Goch High Dam rather than the machinations of the WRSE group. This is robust and future-proofed despite costing an arm and a leg.

I have detailed the relevant proposals elsewhere so will not repeat them.

The Wye

The River Wye runs through Wales and England and has been described as the most beautiful of all of Britain's major rivers. It would be hard to argue with this.

It (or should it be 'she'?) is not heavily used for abstraction and this is probably a good thing as the river has become heavily polluted over the last decade. Despite the machinations of all and sundry, to tell you otherwise, this appears to be the result of building a herd of chicken farms in Powys. If it looks like a duck, walks like a duck and quacks like a duck, then chances are it's a duck! (actually it's chickens and rather - a lot of them).

What environmental impact reports were produced is a mystery and why Natural Resources Wales did not speak up is questionable. Either way, the planning system failed and we now have a farming industry that disposes of its waste in a manner which pollutes the river with phosphates. This results in unsightly and potentially dangerous algal blooms which are destroying the river and its wildlife.

[All of this is a personal opinion, based on what I have read and what some are saying. For a balanced view you should consult with the chicken farmers, the waste disposers and the carcase processors. Other chicken farms are available.]

A suggestion has been made to build a fertilizer plant to process the waste slurry into a pelletized phosphate fertilizer which could be sold. Due to the war in Ukraine, there is a worldwide shortage of such fertilizers at present.

The Thames

Looking at Isis as she flows through Oxford, one is presented with an idyllic picture but the Thames is in crisis due to over-abstraction. This is largely due to the demands of the capital and its population of around seven million. Fortunately the population density is far less upstream but the stress on the catchment is starkly illustrated by the movement of the river's source downstream from Cricklade. Observers argue about which of the sparse springs is now the true source.

The STT would bring some 300Ml/d into the catchment but, if both options were to be implemented, then this could be 600Ml/d. The transfer via the Grand Union Canal is problematic and will not directly benefit the Thames. It might never come to fruition.

The proposal for SESRO is practical and opportune but it is progressing very slowly and not due to come into being until after 2040. If this is to be the destination of water from the STT then it will hold everything up for years. The key question for STT is where and how to feed the water into the catchment? Taking it to SESRO is a no-brainer but this would do nothing for the upper catchment and the tributaries. Consideration needs to be given to feeding the upstream aquifers via irrigation or injection. Feeding directly to farm irrigation systems would reduce the agricultural demand on the whole system and would make sense though the lack of joined-up thinking will mean that it is unlikely to be considered.

We need to consider the whole catchment – not just the aspects which directly affect water supply.

The Trent

The Trent has a base flow which is largely unnatural and, in the past, it was heavily polluted by untreated industrial discharges. Its temperature was also raised as the numerous power stations along its length which used river water for cooling. Both of these problems are largely in the past but the proportion of treated effluent is still high. Birmingham imports over 300Ml/d from Wales before discharging the treated effluent into the catchment via the River Tame. The other substantial effluent discharges largely emanate within the catchment so are not net gains.

The main question relates to whether Trent water can be better used to support resources in the area as the flow is consistent even during extreme conditions. The EA has published its licensing strategy for the Trent but fails to include a description of the existing abstractions. Can we take more raw water from the river or not? The answer is clouded in a plethora of machinations but the answer would appear to be yes. If so, then could Trent water be taken to Rutland Water to support the grid?

There are pressures to reduce the abstractions from the Nottingham sandstone and the Doncaster aquifer and this shortfall needs to come from somewhere else – the Trent is the obvious answer but is the water now good enough to be treated for public supply. Church Wilne is situated alongside the river and has membrane systems installed but this is expensive. Taking shallow ground water from the valley could provide a solution which is both sustainable and of good quality.

Craig Goch and Dol Y Mynach

Plynlymmon in Wales provides the source for three major rivers in North Wales, including the Severn and the Wye. The source of the Dee is not far away. All three have a major part to play in providing water to England and this has been a constant source of discontent to parties in Wales and especially Plaid Cymru.

The Elan Valley dams, at the head of the River Wye catchment, serve Birmingham. The River Severn is regulated by Clywedog and supplies water to communities along its course (see Appendix 2). Lake Vyrnwy is near the head of the Severn catchment and is supported by Llyn Celyn and Lake Bala which also connect with the River Dee.

Proposals to reinforce the Thames catchment, which is in serious deficit, include a proposal to transfer water from the lower Severn into the Thames catchment via Deerhurst. One issue is: how to provide compensating water to the Severn to make up for an abstraction of some 300Ml/d? Proposals to amend the abstraction near Lake Vyrnwy resemble 'smoke and mirrors' and the plan to use Minworth effluent, via the River Avon is likely to provoke objections from the residents of Stratford and Warwick.

There were detailed plans to build the High Dam at Craig Goch in the 1970s and a public enquiry was held. Severn Trent decided, in the face of Welsh Nationalist opposition, to build Carsington Reservoir instead and this was completed in the late 1980s – the last major reservoir to be developed in England.

Craig Goch is the best reservoir site in the whole of England and Wales and, subject to overcoming the political objections, it could provide a resource which would form the basis of a future water grid. It could be developed to provide security to each of the three major rivers in its vicinity. Water from the High Dam could also be gravitated all the way to London.

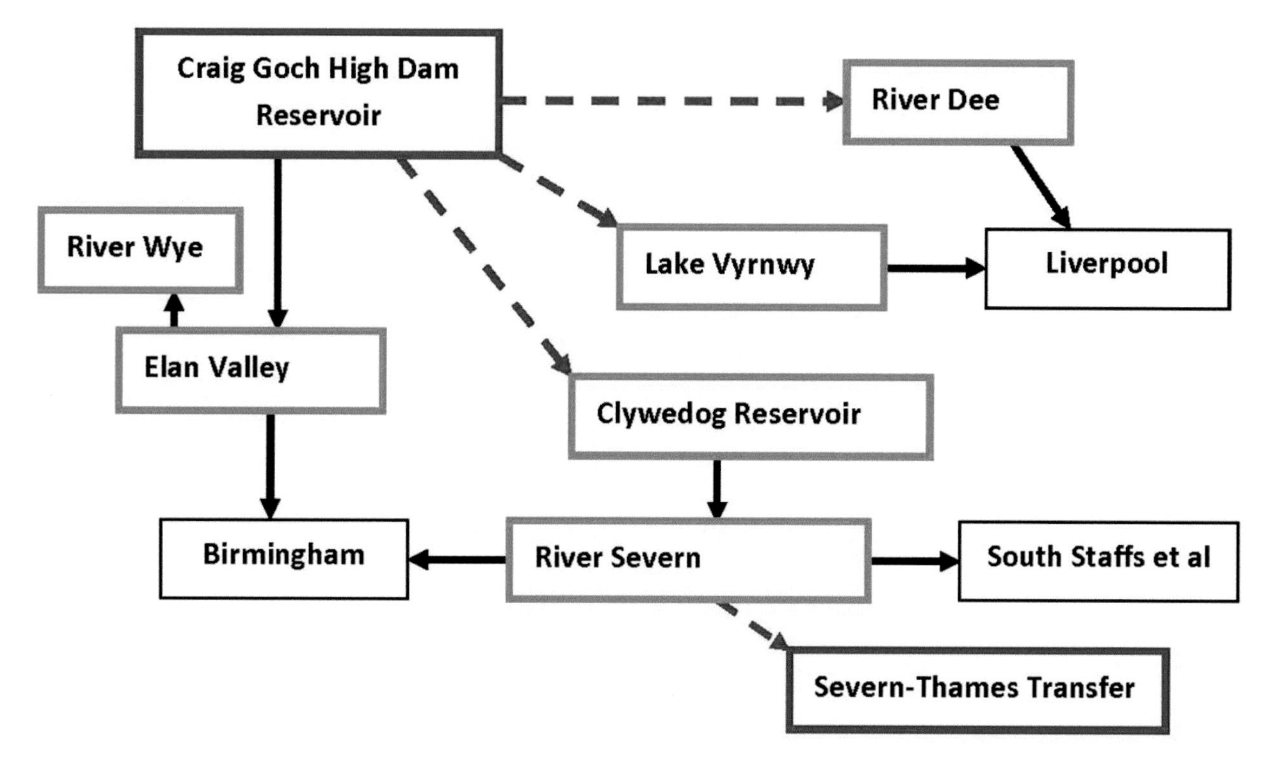

The recent report by the Joint Committee on National Security Strategy (JCNSS) criticised all of those responsible for national infrastructure including the power industry. One of the problems faced in electricity supply is meeting peak demand and there is little infrastructure designed for this especially south of the border with Scotland. The Craig Goch High Dam would be ideally suited to work in conjunction with the lowest of the existing Elan Valley reservoirs, Caban Goch, to provide a pumped storage scheme which would spread the cost of construction over two industries.

An aside:

Dol y Mynach is a part-completed dam which was never finished as Birmingham Corporation decided to build Claerwen instead. The foundation of the dam remains in the forlorn hope that someday it might be needed.

Whilst the original purpose of the dam has been superseded, it could possibly serve another purpose as the foundation was built to support a masonry dam. Central Wales is not a hot-spot for employment and there is a need for trained stone masons across the country. Could a new project involve the development of a masonry training academy based on building the dam? It could be associated with a university and provide high-value apprenticeships for those wishing to make masonry their career. A possible source of income could be to make the dam into a national memorial for those who have been cremated so that an individual stone could be inscribed and placed in the face of the dam.

Just an idea.

Anthony Gostling

The 'Yuck Factor' and Recycling

As water resources become more scarce, many are turning to desalination as a means of treating seawater as an inexhaustible, but expensive, resource. There are a number of such proposals within the reports of the five regional groups and the possibilities of reverse osmosis treatment of treated wastewater i.e. sewage is coming to the fore.

When planning for the 2012 Olympics, Thames Water decided to build a desalination plant on the Thames estuary to treat brackish water to potable standards. This illustrates the issue of the 'yuck factor' perfectly as it would have been easier and cheaper to treat Beckton effluent than brackish, estuarine water. Salt, in solution, is much more difficult to remove than the impurities and organisms which are present in treated effluent from a modern sewage treatment plant.

The key issue is a perception one and the question is a simple one:

"What process or processes does treated effluent have to go through to make it acceptable?"

And the answer, despite all of the complications that surround it is also simple:

"Dilution."

Just look at a map showing the route of the River Severn or the Thames. Each settlement along the route takes water from the river and then treats it before discharging it back to the river. Its discharge consent relates to the environmental impact on the downstream waters and normally a dilution factor of ten to one is achieved. This is accepted throughout the population and never questioned but the direct supply of treated wastewater is still taboo. The plans for the treatment and dilution of water at Havant Thicket reservoir perfectly illustrate this. Sand filtration will normally provide sufficient added treatment where there is little dilution.

Whilst the membrane treatment of waste water is cheaper than using saline waters as a resource, both should be avoided wherever possible due to their cost and carbon footprint rather than the 'yuck factor'.

Invasive Non-Native Species

A water grid which moves treated water about would not risk spreading INNS from one body to another but a raw water grid does carry this risk. Some may consider it to be minimal compared with the risk of water supplies running out but it still needs serious consideration. Measures already exist to identify the risks and treatments are available which should eliminate them. Sand filters will be enough to eliminate most risks. A national database indicated all INNS and their locations would be useful in resource planning.

Aquifers and Leakage

I have referred elsewhere to the issue of leakage and wish to reinforce my view that leakage reduction is largely irrelevant in overall resource management terms. A recent paper published by one of the major water companies showed their strategy for dealing with the projected resource shortfall over the next decade – a requirement set by Ofwat – to be a reduction in leakage.

The initial shortfall is to be met by the development of more boreholes and long-term needs by leakage reduction. Actually, neither of the these affects the overall availability of water resources in an area, whether it be big or small. Boreholes simply deplete an aquifer (as we have seen in the South of England) and leakage recharges it.

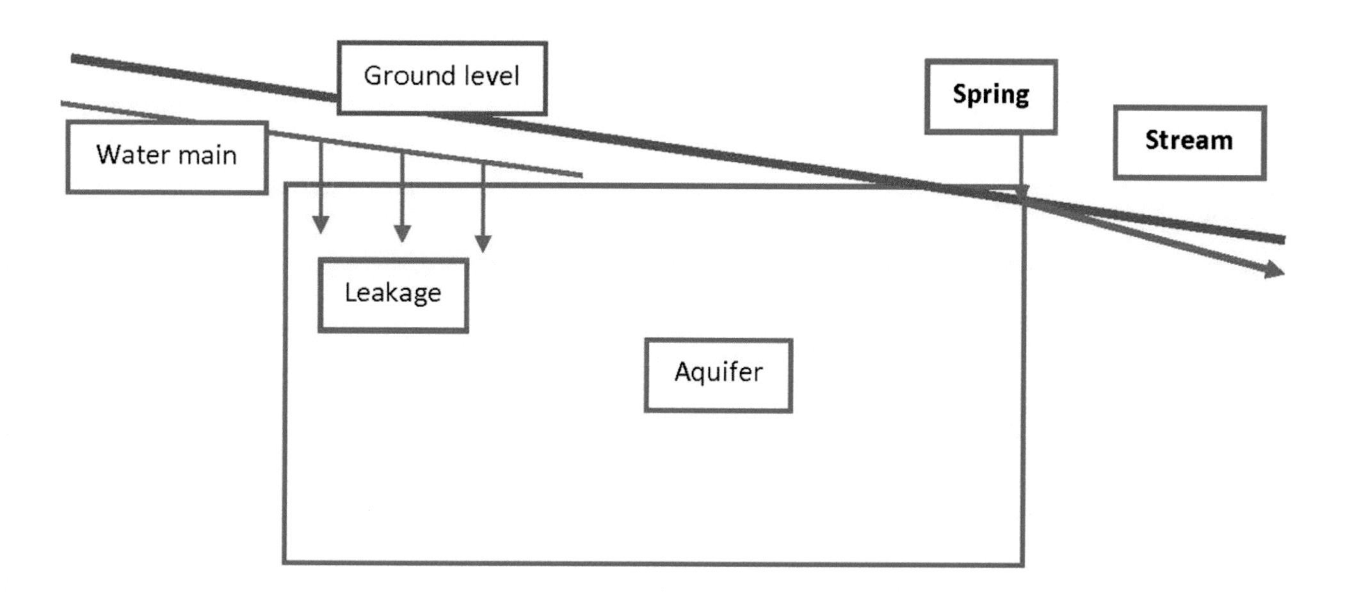

This is not rocket science – it's common sense – but water professionals continue to think that reducing leaks has a beneficial effect on reducing demand on the aquifers and watercourses. Many minor watercourses. in towns and cities, would dry up complexly in dry weather if it were not for leakage from the water mains. I heard it said, though I can't prove it, that the River Cole, which drains part of Birmingham, has a background fluoride level similar to that in the water supply. Guess where it comes from?

In London, and other coastal cities, the picture is a little different as the local aquifer is likely to connect with the sea or brackish waters. In this case, leakage into the aquifer helps to hold back saline intrusion.

Agriculture and Irrigation

The need for water in agriculture is almost as important as that for public supply however there are differences. Most of the abstraction for crop growing is seasonal and short term whilst horticultural demand can be all year round.

This means that agricultural water is often required when the whole system is already stressed. In addition it is self contained within the catchment if it is returned as irrigation water which flows to the local shallow aquifer and hence does not greatly deplete the overall resource.

The abstraction may be short-term but the resulting recharge is long-term as it will take the irrigation water some time to get through to the aquifer. Spray irrigation is inefficient and results in heavy losses due to evaporation. Drip irrigation is efficient but can only be used where there is no need for ploughing. Other losses include process water and that exported in the crop.

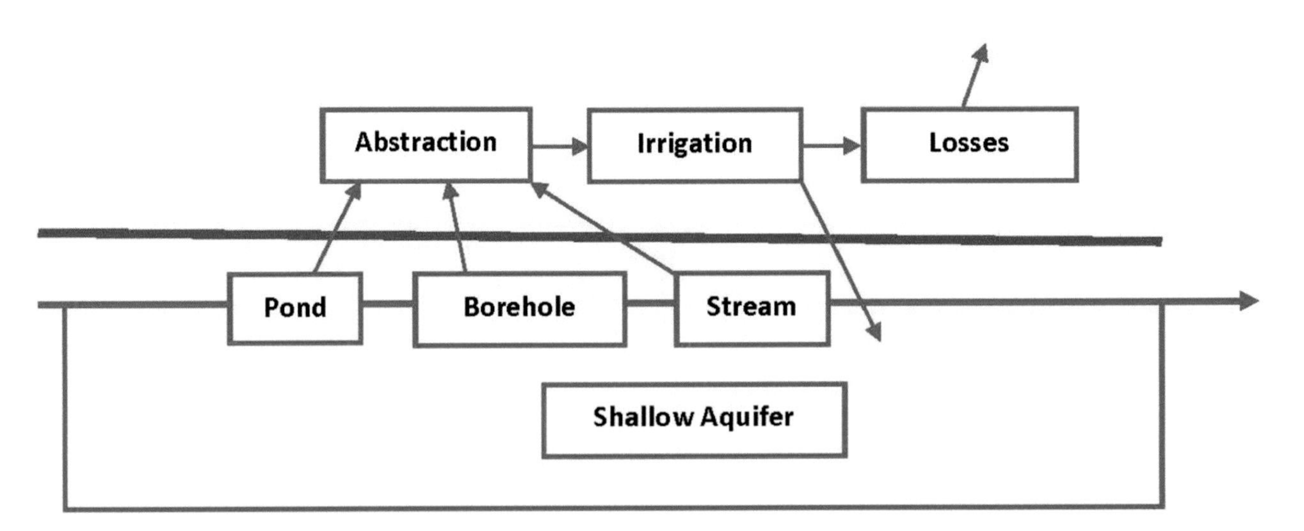

Obviously, the needs of agriculture need to be taken into account in the overall strategy and the NFU have plans to draw up their own resource plans alongside those currently being developed. However, it is clear that, in overall resource terms, that any net abstraction is a draw on finite resources and hence a reduction in that available in the long term.

If we consider the STT supply to be about 300Ml/d then there are three obvious ways it could be added to the Thames catchment:

- A pipeline direct to SESRO (say 100Ml/d)
- Direct addition to streams and rivers (say 100Ml/d) and
- Drip irrigation to the shallow aquifers in the uplands (say 100Ml/d

Drip irrigation in the uplands is the most eco-friendly of these options and could be useful in restoring the chalk streams as well as meeting agricultural needs. Obviously, if the upstream aquifers are healthy, then wetlands and habitats will flourish. In the same way, any upstream work to reduce flooding will benefit the overall resources in the catchment.

Water is not intelligent – it doesn't know why it resides in a catchment or an aquifer or why it's being abstracted. So, when we import water to a catchment, it's not relevant whether it is brought in by a water company or another body – it's an import – full stop.

Virtually all hot countries have irrigation systems which are usually constructed at government expense and run by collectives – much like our Internal Drainage Boards but with the opposite purpose. Farmers make demands on water resources during dry periods and this adds to any deficit resulting from abstraction for water supply. It doesn't matter why the water is taken – it all adds up.

The failure of relevant bodies to comprehend this just adds to the silo mentality which I have moaned about already. Whilst I was no fan of MAFF, they did at least understand that irrigation was just as important as drainage. Moving on, we need to take the needs of agriculture on board in any plans to move water around the country and thus involve the NFU as a partner in discussions.

The health of a catchment is dependent upon the health of its aquifers and spray irrigation, in the absence of rainfall, is one method of returning an aquifer to its healthy state. This process imitates sand filters and thus should guard against the import of INNS. Drip irrigation is more efficient but has limitations.

Asset Management

One of the key components of regulation is asset management and the regulator runs a five year programme against which the water companies are measured. Many argue that this is responsible for the short-term thinking that pervades the industry and the 'peaks and troughs' in the resulting capital programmes,

There is, however, another aspect that the industry should pursue. Whilst each County Council holds a database of reservoirs in its area for safety purposes, I am not aware of any such holding in relation to water resources. I would recommend that a national database of reservoirs be compiled to store basic information, including safety inspections, but also to include performance data.

For example:

- Name
- Location
- Owner
- Operator
- Type
- Description
- Size
- Catchment
- Interlinking
- Purpose/supply/regulation
- Inspections
- Condition grades
- Performance grades

Obviously condition grades would come from the safety inspections but the latter issue is a new facet. This should describe, using the standard five grades of asset management, what resilience the reservoir has to periods of drought. For instance:

1. Will withstand a 1000 year drought
2. Will withstand a 100 year drought with spare capacity
3. Will withstand a 50 year drought
4. Will fail in a 50 year drought
5. Will fail in a 20 year drought

I am not aware of any industry standard in this respect but suggestions can be gleaned from *Principles of Asset Management* by Styles and Earp.

Funding

It could be, given the current mode of regulation with the privatised water companies, that funding is a major obstacle to the development of a National Water Grid. The Board of a water company, seeing their responsibility to shareholders, will be reluctant to be the first to suggest a transfer from outside lest they be required to fund it.

If water is to be transferred between one region and another, there needs to be a clear understanding of what the funding mechanism will be. There are two issues to be considered:

1. Who pays for the initial investment in building the infrastructure?
2. How much is to be paid for the transferred water and by whom?

There are a number of historical examples but privatisation and the Welsh Nationalists caused a few things to be amended. The normal rule is that the receiving body (say Birmingham Corporation Waterworks) pays for the land and to build the dams and the pipelines i.e. the infrastructure and then they own it. Birmingham got a big surprise when the Elan Valley was transferred wholesale to Welsh Water on privatisation and they took the government to the high court – where they lost as the transfer was enshrined in the enabling act.

No payment was made for the exported water from Wales to Birmingham or Liverpool. Protests by the Welsh Nationalists resulted in Severn Trent setting up a trust fund to pay for the export and there is an on-going arrangement though largely related to operation and maintenance. When a senior union member suggested that Craig Goch should be enlarged, Plaid Cymru objected in no uncertain terms resulting in the withdrawal of the comments. It is clear that a beneficial funding arrangement needs to be made for any water exported from Wales to England*.

The obvious solution would be for government to fund any construction works on the basis that it is in respect of the national interest. Maintenance could be a similar charge against the public purse. The operating costs of the pumps etc. would need to be covered and this would normally be by the body who receive the water.

Overall, such issues remain problematic so long as there is no clear owner of the problem. This is a matter for Government to resolve.

*300Ml/d at 20p per m3 would come to roughly the same as the Chief Exec's basic salary.

Organisation

If we are to take resilience seriously – and there's little evidence to date that we are – there needs to be a driving force at the political level to make things happen. The current system, whereby Ofwat strictly controls expenditure within the companies' boundaries, will never bring about change. Likewise, the Environment Agency's half-hearted attempts to drive things forward have come to little.

As recommended by the Joint Committee report, we need someone to take overall control at ministerial level, reporting to the Secretary of State responsible for the environment. In order to avoid undue bureaucracy, a small team would oversee things and make use of contracts for services.

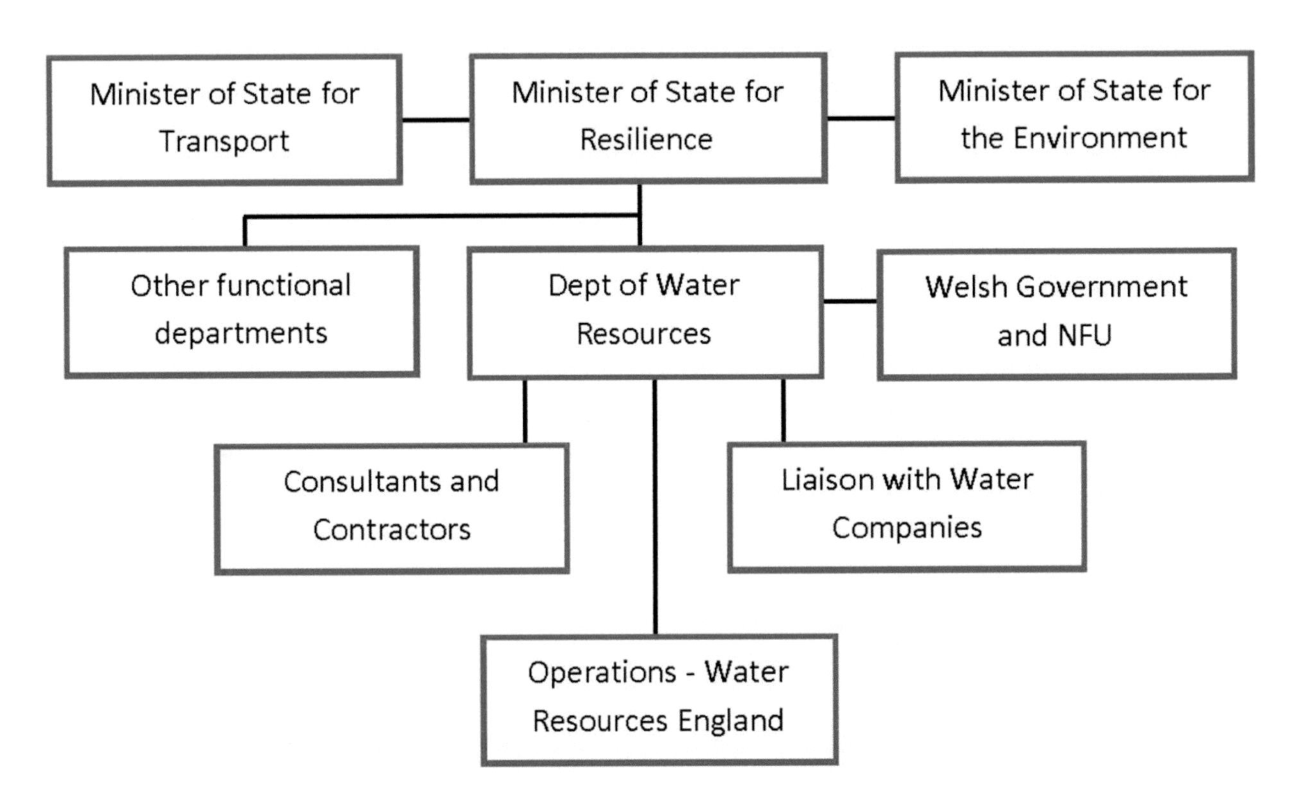

A question remains – how would a grid work on a day-to-day basis? Would there be a central control or would the links be run by the local water companies in consultation with their neighbours? Some sort of overall monitoring is essential but would it also have direct control or simply request the local company to make a transfer?

Would the five regional groups have any part in this? On the plus side we could get rid of five acronyms and replace them with just one (WRE). Make your own mind up.

Conclusions

Steps need to be taken in the face of climate change to ensure that England's water resources are secure. This is unlikely to happen with the current model for the industry which ensures that water companies operate largely in isolation from each other. A first step to resolve this would be to amalgamate the water-only companies into the ten regional companies using a generous share offer underpinned by government.

A top-down approach is needed in order to break away from the current bottom-up, silo mentality that pervades the industry. The current approach with the five regional groups, overseen by the EA, is only working in a local context.

As recommended by the JCNSS, a government minister needs to have responsibility for ensuring that utilities are made resilient from climate change and other threats. Government needs to take responsibility for much of the funding.

Leakage and consumption reductions will have little effect overall despite what the regulators and the companies are saying.

A raw water grid, alongside new reservoirs, will provide the best defence against future raw water shortages.

The scheme to build the High Dam at Craig Goch should be resurrected with a view to developing both water resources and a pumped-storage facility.

A new financial model for the funding of inter-regional water transfers is needed, and especially in respect of Welsh water.

My initial proposal for a grid in the Midlands and the South was published in 2014. If it had been actioned <u>then</u>, we would be building the first links <u>now</u>.

A closing thought

Unfortunately, our agenda will not be set as a result of forward planning and building for the future. It will be determined by television images when disaster strikes and the taps run dry. A blame game will then ensue, heads will roll and "lessons will be learned". But once the disaster has passed, everyone and his dog will go back to sleep and we will be left with little to pass on to the next generation.

The Victorians did not behave like this and neither should we. So let us join up the dots!

[the following image is courtesy of Portable Antiquities via Openverse and Creative Commons]

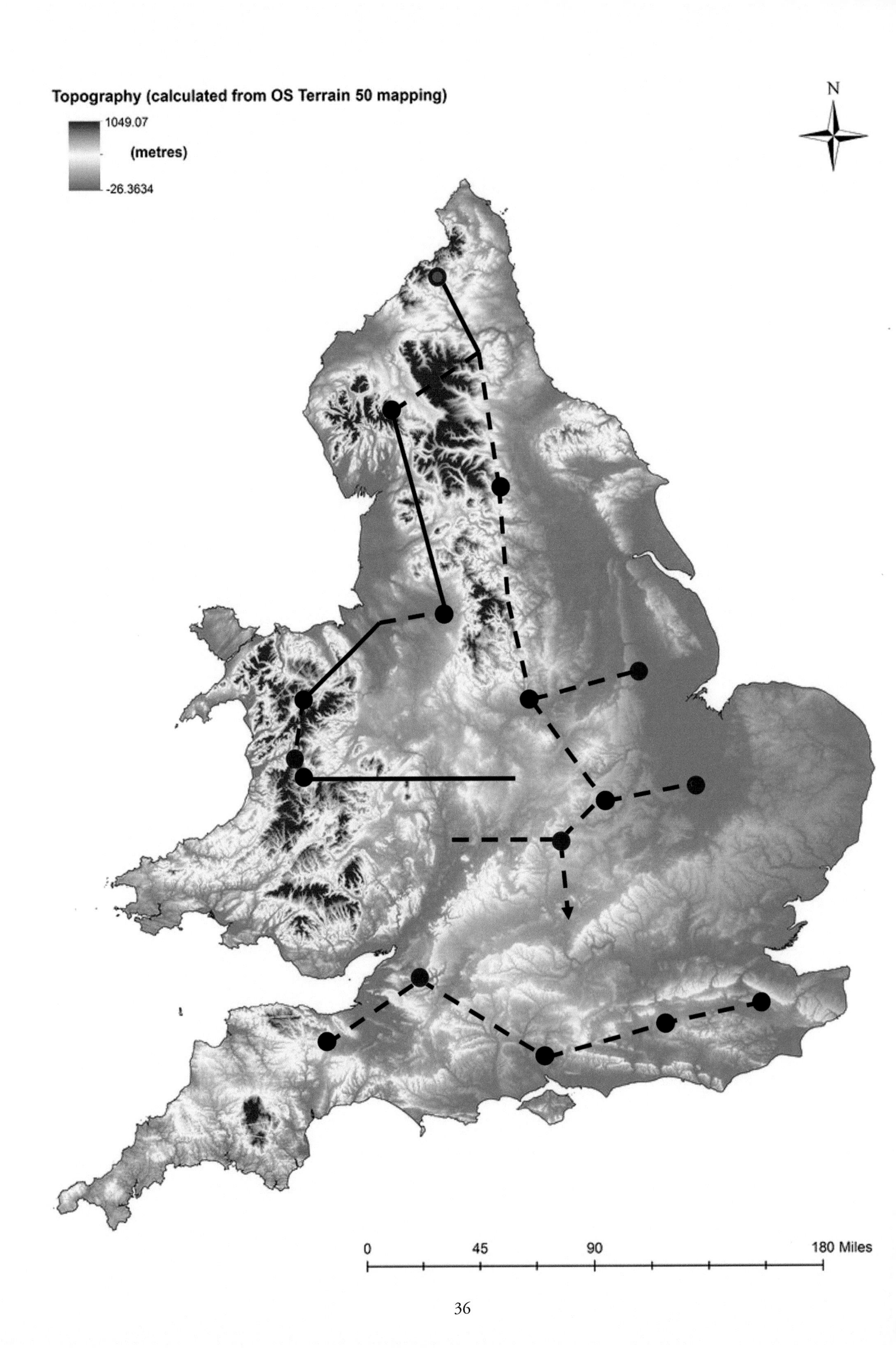
Topography (calculated from OS Terrain 50 mapping)
1049.07
(metres)
-26.3634
N
0
45
90
180 Miles

Appendix 1: Regional Groups' Terms of Reference and Extracts

[From 'Meeting our future water needs: a national framework for water resources' in 16 March 2020]

"Regional groups should:• scope a wide range of supply options, such as reservoirs, water reuse and desalination, with a clear understanding of how long each would take to be implemented to allow options to be brought forward if demand is not reduced as expected

- explore the strategic options funded as part of Ofwat's gated process
- identify new options that are not included in the current plans and engage in the catchment based approach, particularly in priority catchments, to develop cross-sector options that provide broader benefits to society
- **investigate the potential for increasing connectivity within and between regions through:**
  - **longer distance transfers, such as those over 100km in length, and those that also include water storage to increase drought resilience**
  - **shorter transfers that increase resilience to interruptions in supply**
- When exploring transfers regional groups should:
  - consider the potential to make them reversible so that they can increase the resilience of both parties
  - be clear on how transfers would be used during droughts, including when one or both supplier or receiver is implementing drought management tools
  - work with the Environment Agency, Drinking Water Inspectorate (DWI) and the Regulators Alliance for Progressing Infrastructure development (RAPID) to make sure that planned transfers are feasible and that any issues are carefully managed"

Extract 1

Senior Steering Group (June 2022)

"The meeting agenda was shaped to help manage the following risks from the risk log maintained by the regional coordination group:

- The risk that there is a lack of alignment between regional plans – in particular the risk that strategic transfer schemes do not line up between plans

- The risk that, despite investment in managing demand, it is not possible to reduce consumption to the levels water companies and regional groups are aiming for. This could have substantial impact on the plans which are very sensitive to projected demand."

Extract 2

RAPID expects (Oct 2020) Plans that:

"Are ambitious on the environment, including changes to water abstractions, reducing the use of drought orders and permits, and exploring opportunities to deliver wider environmental benefits.

- Make progress on water demand management by striving to achieve per capita consumption of, on average, 110 litres per person per day by 2050 while also reducing non-household demand and delivering the industry's target to reduce leakage by 50% by 2050.

- Explore a wide range of options to increase water availability, including reservoirs, transfers, water reuse schemes and desalination plants and arrive at a plan that delivers for your customers, your region and the nation.

- Identify opportunities to develop and share supplies with stakeholders in other sectors and deliver broader social benefits such as reducing flood risk."

Extract 3

"The Environment Agency estimates that if no action is taken between 2025 and 2050 around 3,435 million extra litres of water per day will be needed for public water supply to address future pressures. But parts of England are forecast to run out of water in the next 20 years, and our available UK water supply is forecast to drop 7% by 2045 due to climate change and sustainable abstraction limits.

The Environment Agency intends that the plans will:

- Reduce water demand, per person and across sectors.
- Halve leakage rates by 2050.
- Develop new water supplies, such as reservoirs and reuse schemes.
- Move water to where it's needed.
- Reduce the use of drought measures that impact the environment."

Appendix 2: The River Severn

The River Severn as a Water Resource (extracted from *The Power of Water*).

The River Severn (or Sabrina/Hafren as it is sometimes known when endowed with character) is not just a land drainage channel as it also serves as a navigable waterway and a source of raw water for the communities along its route. It's use as a water resource has developed considerably over the years and it is now managed to a very great extent. The weirs, which were built to aid navigation, also serve as devices to measure the flow. Strangely, Birmingham gets most of its water from the adjoining River Wye catchment due to the City fathers' decision to develop the Elan Valley dams. Water quality (at that time) and the lack of pumping would have contributed to their decision making. Trimpley was built much later as a back-up to the Elan Valley supply. The Severn is now navigable as far up as Stourport-on-Severn but, in the past, boats went as far north as Bewdley.

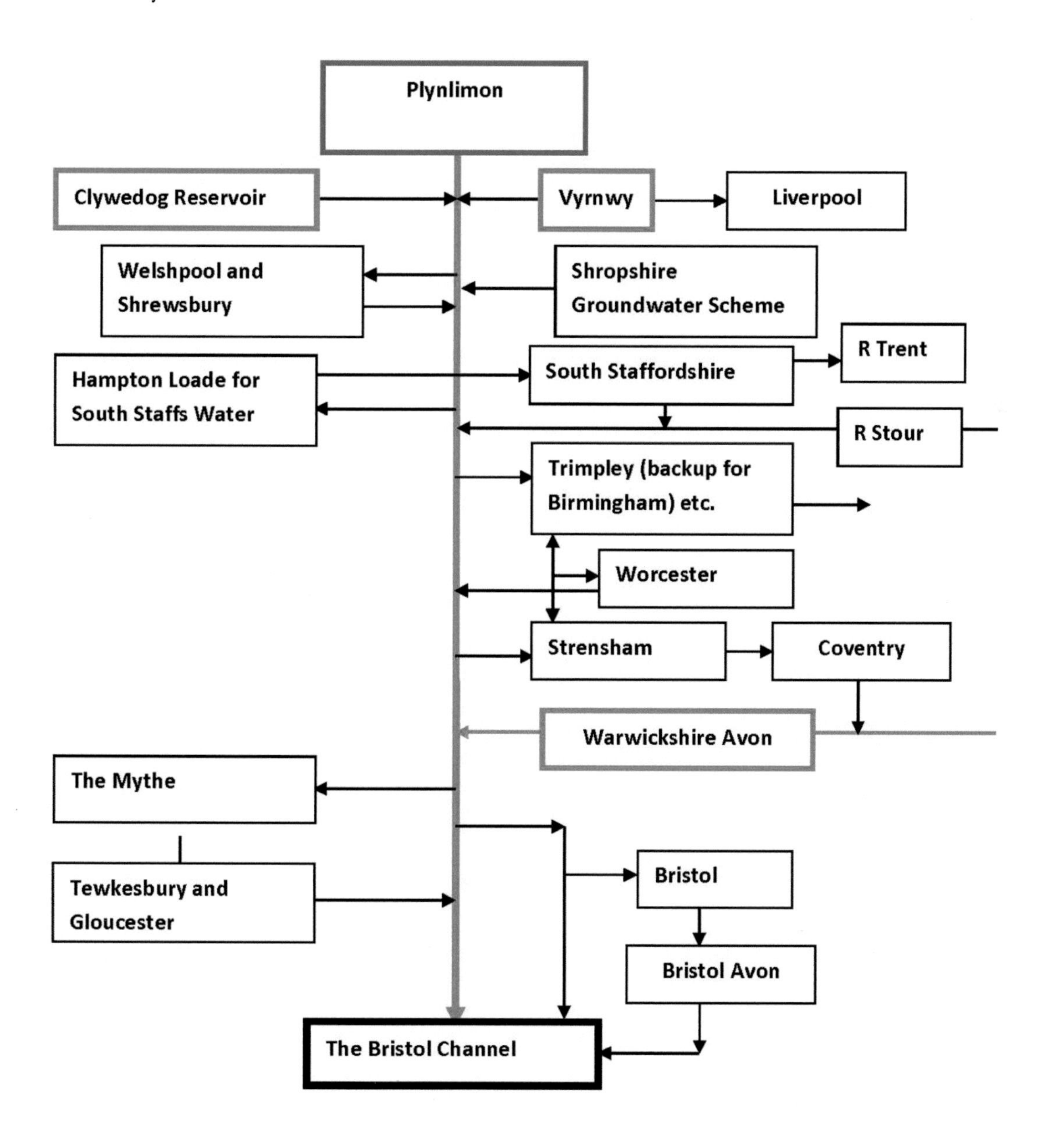

Appendix 3: A Water Grid for England

This 'paper' is based on a submission to ICE following a consultation on the future of water resources in 2014.

Most water companies have an internal 'strategic grid' which enables the major water treatment plants to support eachother when there are problems. However, these grids are largely parochial in nature and, since the demise of the Water Resources Board, joined-up thinking has been in short supply. Once mention is made of a 'water grid' everyone and his dog come out to say that it's impossible and too expensive. Actually a substantive grid, moving raw water around between existing major reservoirs is both practical and economic – the likely overall cost being less than is being spent on the Thames Tideway project.

The key issues are:

- Only transfer raw water and
- Make optimal use of existing infrastructure

Overlaid on a diagram of the existing sources and assets, the proposed links (in red) shown below could form the basis of an affordable water grid which would solve much, though not all of the problem. The key components involve the interlinking of the North Midlands systems with the reservoirs which serve much of the Anglian region. Taking water from the Trent is dependent upon quality issues though these are not insurmountable especially when the choice is between providing expensive treatment or running out of water to treat. Transfers into the Thames catchment can be made through an extension of this system, from the Warwickshire Avon, from the Bristol Avon, from the Severn and even from the lower reaches of the Wye.

The areas south of the Thames and Kent are shown in the second diagram. This requires more lateral thinking than simply interlinking the raw water reservoirs. Kent and the South are inherently in deficit and the situation is not helped by the plethora of small private water companies.

Interlinking of the Kent reservoirs along similar principles to those suggested in the Midlands is a simple matter of geometry but the provision of additional resources is not. Importing water by tanker from the North East (Kielder) can be achieved by unloading at Gravesend to Bewl Water and at Southampton for Testwood Lakes. In addition, it should be possible to provide additional treatment for effluent from Crossness and transfer this for blending in Bewl Water.

Missing from the diagram are the systems bringing water from the Lake District to Manchester, and from Kielder Water to support the rivers south of there. Also Ladybower serves both Severn Trent and Sheffield under an agreement which runs out in 2030.

It took over ten years to interlink the East and West Midlands following the 1976 drought and twenty years for Severn Trent to complete its strategic grid. How long will this take?

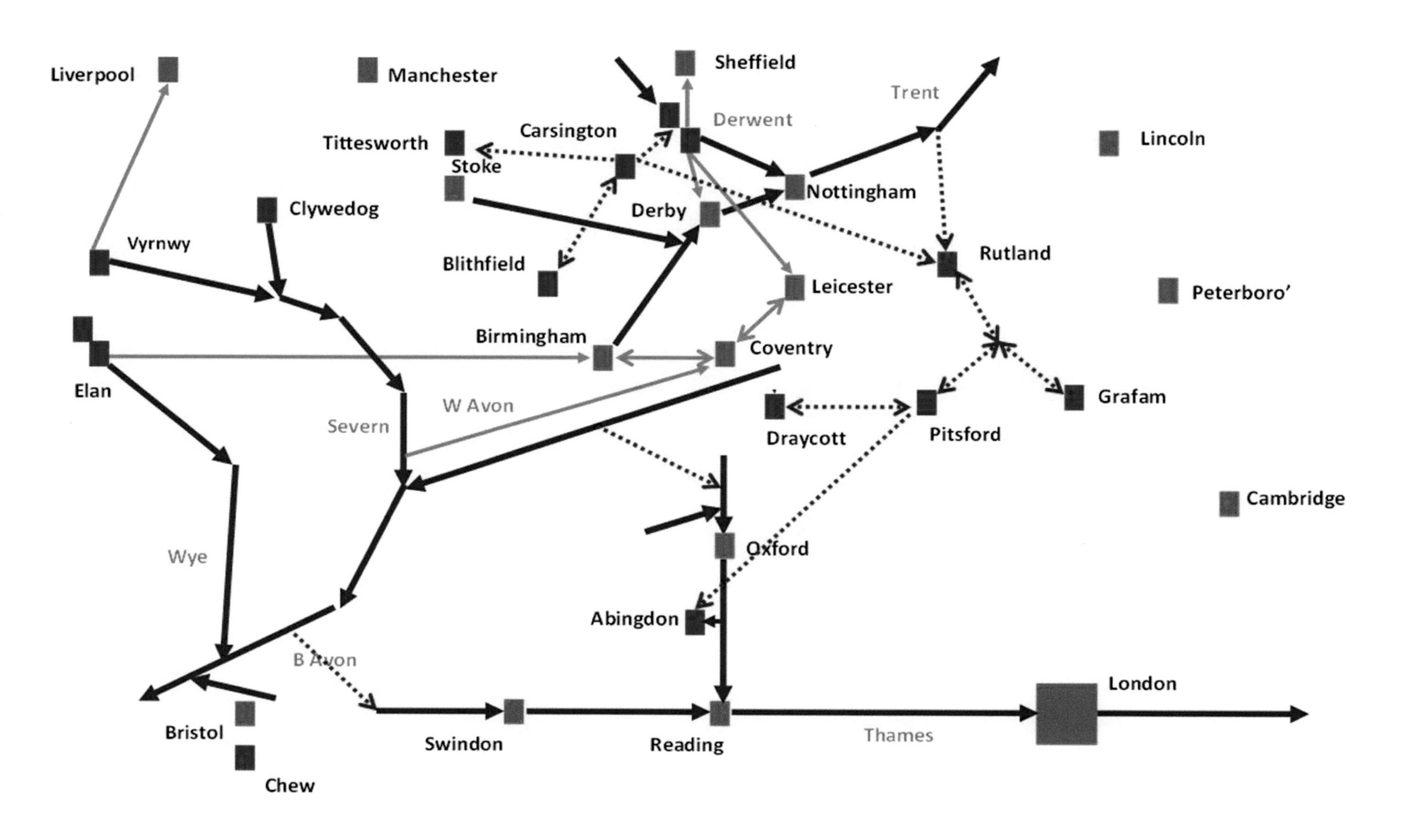

Liverpool
Manchester
Sheffield
Trent
Derwent
Carsington
Tittesworth
Stoke
Lincoln
Nottingham
Clywedog
Derby
Vyrnwy
Blithfield
Rutland
Leicester
Peterboro'
Birmingham
Coventry
Elan
Grafam
W Avon
Severn
Draycott
Pitsford
Cambridge
Wye
Oxford
Abingdon
B Avon
London
Bristol
Swindon
Reading
Thames
Chew

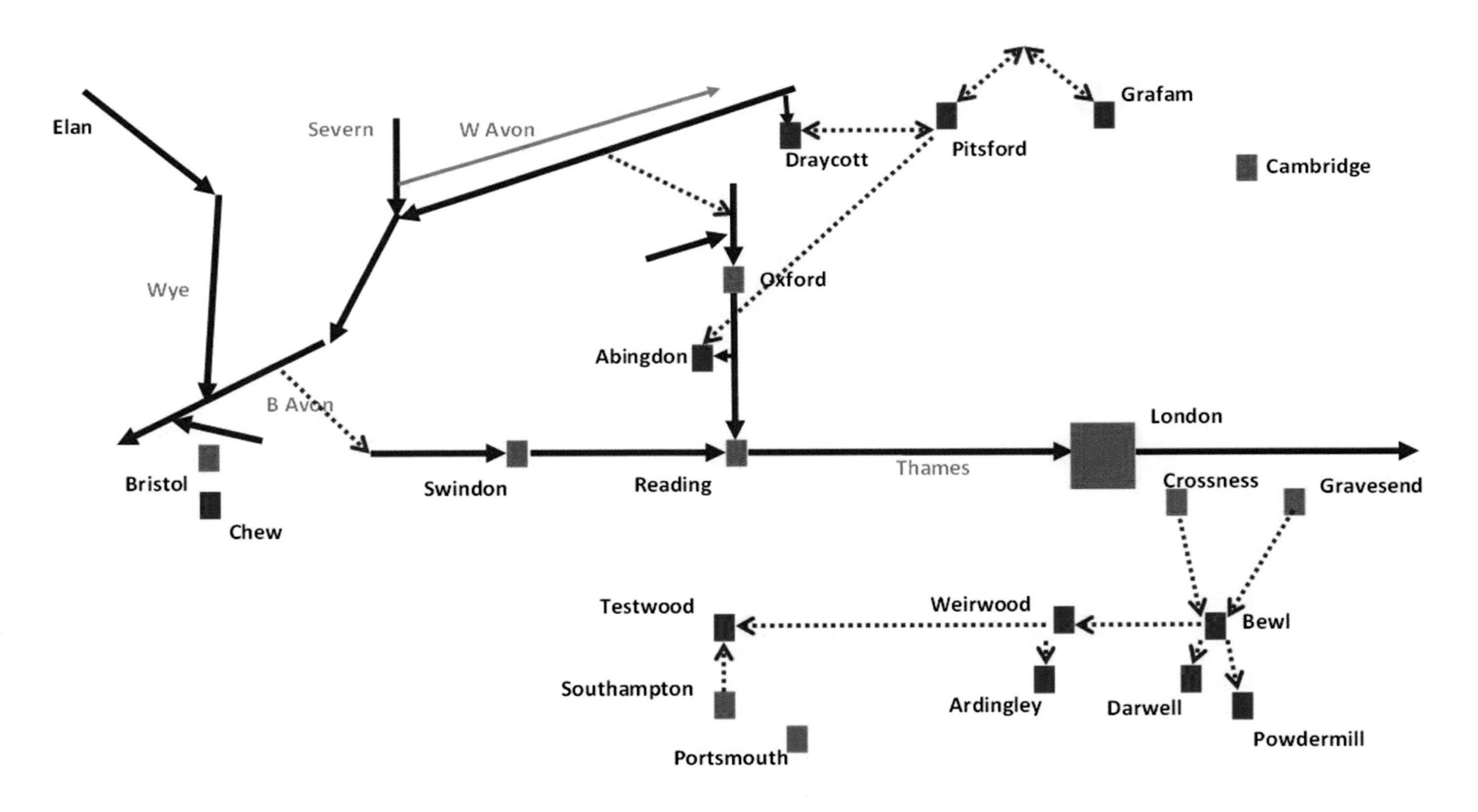
Elan
Severn
W Avon
Draycott
Pitsford
Grafam
Cambridge
Wye
Oxford
Abingdon
B Avon
London
Bristol
Swindon
Reading
Thames
Crossness
Gravesend
Chew
Testwood
Weirwood
Bewl
Southampton
Ardingley
Darwell
Powdermill
Portsmouth